Essential Fluid, Electrolyte and pH Homeostasis

Gillian Cockerill

Reader, Vascular Biology, St George's London

Stephen Reed

Principal Lecturer, School of Life Sciences, University of Westminster

WILEY-BLACKWELL

A John Wiley & Sons, Ltd., Publication

This edition first published 2012

Wiley-Blackwell is an imprint of John Wiley & Sons, formed by the merger of Wiley's global Scientific, Technical and Medical business with Blackwell Publishing.

Registered Office
John Wiley & Sons Ltd, The Atrium, Southern Gate, Chichester, West Sussex, PO19 8SQ, UK

Editorial Offices
350 Main Street, Malden, MA 02148-5020, USA
The Atrium, Southern Gate, Chichester, West Sussex, PO19 8SQ, UK
9600 Garsington Road, Oxford, OX4 2DQ, UK

For details of our global editorial offices, for customer services and for information about how to apply for permission to reuse the copyright material in this book please see our website at www.wiley.com/wiley-blackwell.

Library of Congress Cataloging-in-Publication Data

Cockerill, Gillian.
Essential fluid, electrolyte, and pH homeostasis / Gillian Cockerill, Stephen Reed.
p.; cm.
Includes index.
ISBN 978-0-470-68306-4 (pbk.)
1. Body fluids. 2. Acid-Base Equilibrium. 3. Water-electrolyte balance (Physiology) I. Reed, Stephen, 1954- II. Title.
[DNLM: 1. Body Fluids–chemistry. 2. Acid-Base equilibrium–physiology. 3. Body Fluid Compartments–physiology. 4. Water-Electrolyte Balance–physiology. 5. Water-Electrolyte Imbalance. QU 105]
QP90.5.C55 2011
612′.01522–dc23

2011015327

A catalogue record for this book is available from the British Library.

This book is published in the following electronic formats: ePDF 9781119971894; ePub 9781119973669; Wiley Online Library 9780470683064; Mobi 9781119973676

Set in 9/11pt Sabon by Aptara Inc., New Delhi, India.
Printed and bound in Singapore by Fabulous Printers Pte Ltd
First 2012

In loving memory of
MW, EMS, IW and JFH

Contents

Part 2: Fluid and electrolyte homeostasis

Normal physiology

List of figures

List of tables

Preface

> 'All vital mechanisms, no matter how varied they may be, have always but one end, that of preserving the constancy of the conditions of the internal environment.'

With these words, Claude Bernard in 1857 gave what was probably the first definition of the process we now know as 'homeostasis', even though the term, derived from the Greek *homoios* meaning 'the same', was not coined until the 1920s.

Those events that collectively constitute 'life' can all be described in terms of chemical and physical processes; cell biology in particular is chemistry in disguise! The purpose of this short text is to act as a primer for students meeting key topics for the first time, but sections of this book will also be useful as a quick revision guide for more advanced students. The text, supported by diagrams, aims to explain physiochemical processes related to the homeostatic maintenance of:

(i) electrochemical neutrality (anion/cation balance);
(ii) osmotic balance (regulation of the concentrations of solutes inside and outside cells, and
(iii) hydrogen ion balance.

The mechanisms of fluid, electrolyte and acid-base homeostasis are fundamental to normal cellular function and therefore have a major impact on the health of the individual, and an imbalance may lead to a life-threatening situation. Processes of fluid, electrolyte and acid-base regulation that are physiologically interrelated are the ones which students often find most difficult to understand, partly because of their complexity.

This introductory text is divided into three main Parts dealing initially with basic physicochemical concepts, then aspects of normal and abnormal physiology. Each part is presented as a number of Sections which are essentially 'bite-sized chunks' of key information. The book is designed such that it may be read as continuous prose, or, and because each Section more or less stands alone, the reader may dip into the text for the purposes of review or revision of particular topics. Some concepts are described in several sections to ensure that relevant sections are fairly self-contained, but will also allow the reader the opportunity to revisit and consolidate essential material. The contents covered range from basic chemistry and physiology to more advanced concepts which are applied to clinically relevant situations. Selected aspects of analysis and discussion of some of the pitfalls of interpretation of laboratory data are also to be found. There are numerous Self Assessment Exercises based on understanding of key concepts, data-handling problems and case studies for reinforcement of the learning process. We hope the text will be of value to laboratory staff and ward-based staff in endeavouring to understand what many see as a 'very difficult' topic area.

Acknowledgements

Our thanks go to the production and editorial staff at John Wiley; Nicky McGirr, Fiona Woods, Izzy Canning and Celia Carden, Liz Renwick and Samantha Jones. To several colleagues especially, David Gaze and Dr Nawaf Al-Subaie (both St George's) for help with case studies, and Alison Boydell (Westminster) who offered constructive comment on the manuscript.

Also, to Colin Samuell, a valued colleague who provided some of the case histories, and more importantly an inspiring teacher who has made many difficult concepts understandable to countless numbers of students (including SR).

PART 1

Background theory and basic concepts

Overview

The purpose of Part 1 is to review some important concepts of physical chemistry and to introduce key ideas of physiology, all of which will provide underpinning knowledge for deeper study in Parts 2 and 3. Although some understanding of solutions, acids, bases, pH and buffers may have been acquired from previous studies, these topics are included here for revision; some readers may choose to omit certain sections.

An overview is given of body fluid compartments, their volumes and their chemical compositions. Importantly, concepts relating to osmotic balance and electrical neutrality of physiological fluids are also discussed.

Essential Fluid, Electrolyte and pH Homeostasis, First Edition. Gillian Cockerill and Stephen Reed.

SECTION 1.i

Introduction and overview

The human body is, by weight, predominantly water: the total volume[1] being distributed into two major compartments. The larger proportion is located inside cells (intracellular fluid, ICF) with a smaller volume occurring as extracellular fluid (ECF). To function effectively, cells must maintain correct fluid volume balance, ionic balance, osmotic balance and acid-base balance. Two fundamental physicochemical phenomena, namely electroneutrality and osmosis ('osmoneutrality'), have significant effects on cellular function. Homeostatic mechanisms operate to maintain physiological steady-state conditions of ionic and solute concentrations.

Body fluids are complex 'cocktails' of various chemicals such as (a) ions (electrolytes), notably sodium, potassium, calcium, chloride, phosphate and bicarbonate, (b) small molecular weight metabolites such as glucose, urea, urate (uric acid) and creatinine, and (c) larger molecular weight compounds, for example, proteins and lipoprotein complexes.

Qualitatively, the chemical composition of most body fluids is substantially the same, but quantitatively, the chemical content of the different body fluids varies considerably. *Overall*, the total volume of water in our bodies does not vary greatly, and nor does the *overall* chemical composition of the fluids, as both volume and composition are carefully regulated to maintain homeostasis. However, as is often the case in physiological systems, there is at the molecular level a

[1]One should always refer to fluids (i.e. liquids and gases) in terms of *volume* rather than amount.

Essential Fluid, Electrolyte and pH Homeostasis, First Edition. Gillian Cockerill and Stephen Reed.

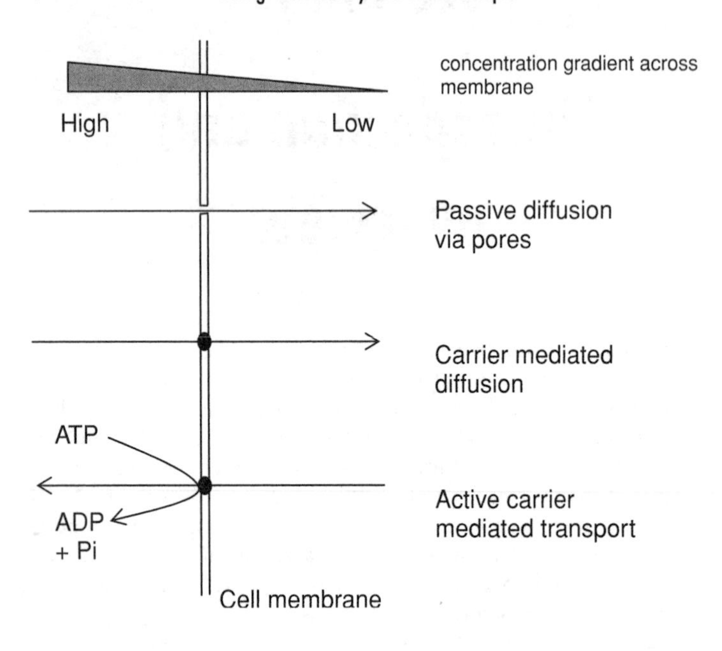

Figure 1.1 Membrane transport. Passive osmotic, passive mediated and active mediated mechanisms

dynamic state of flux and continual change occurring, with fluid and solutes moving between compartments. These movements are driven by physicochemical gradients which arise due to osmotic, electrochemical and concentration[2] differences across cell membranes. Fluid movement between the intracellular and extracellular compartments is directed by osmosis (a particular form of passive diffusion), but because the outer plasma membrane of cells is relatively impermeable to most solutes, especially ions, active or passive carrier mechanisms are required to transport such components between compartments.

[2] Strictly speaking, it is thermodynamic 'activity' rather than concentration gradients which determine the dynamic flux of molecules across a membrane.

Electrochemical gradients *across* cell membranes arise due to the number and nature (principally their size and charge) of the solutes distributed on either side of the membrane. An imbalance in electrical charge across certain membranes is physiologically essential for example to allow nerve impulse conduction and for the initiation of muscle contraction, for example. However, for most cells, an equal distribution of total number of anions and cations is the norm. In addition to the necessity for electrical neutrality *across* a membrane, i.e. between compartments, the numbers of negative and positive charges *within* a particular compartment or body fluid must also be equivalent.

Normal hydrogen ion concentration in most body fluids is very low, in the nanomolar range, compared with concentrations of other ions such as sodium and potassium which are present at millimolar concentrations. Homeostatic mechanisms that regulate hydrogen ion balance are of necessity very sensitive to avoid the wide fluctuations in pH which might seriously impair enzyme function, leading to consequent cell dysfunction.

Physiological control of body fluid volumes and their chemical composition is fundamental to the health and wellbeing of cells, tissues and whole organisms, and as such represents a major purpose of homeostasis. Several physiological systems are involved with normal fluid and electrolyte homeostasis, and thus disorders of the kidneys, liver, endocrine system and gut can all lead to fluid, electrolyte or pH imbalance.

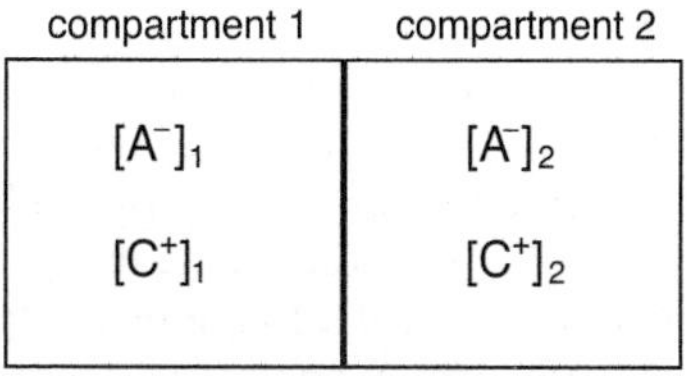

compartment 1	compartment 2
$[A^-]_1$	$[A^-]_2$
$[C^+]_1$	$[C^+]_2$

Electrical Neutrality.

The numerical product of the cations and ions in compartment 1 = the product of cations and anions in compartment 2; $[C^+]_1 \times [A^-]_1 = [C^+]_2 \times [A^-]_2$

Total cation concentration in compartment 1 = Total anion concentration in compartment 1, $([C^+]_1) = ([A^-]_1)$. Similarly for compartment 2.

The specific nature of anions and cations in the two compartments may differ.

Figure 1.2 Ionic balance within and between compartments

The organs that play the most significant role in fluid and electrolyte homeostasis are the kidneys, which process approximately 140 litres of fluid containing a significant quantity of solutes, including electrolytes, each day. Thus, it is not surprising that renal disease is often associated with serious fluid and electrolyte imbalances. Mechanisms for the normal renal handling of water, sodium and potassium, and calcium, in particular, are controlled by the endocrine system via the actions of anti-diuretic hormone (also called vasopressin), aldosterone and parathyroid hormone respectively. Endocrinopathy of the pituitary, adrenal cortex or parathyroid glands will therefore result in fluid and electrolyte imbalance. Renal regulation of proton excretion and bicarbonate reabsorption is a crucial aspect of pH homeostasis but, in contrast to water and electrolyte balance, is not under hormonal control.

The gastrointestinal tract secretes and subsequently reabsorbs a large volume of electrolyte-rich fluid on a daily basis, hence conditions leading to severe vomiting or diarrhoea can also result in significant losses of water, electrolytes or acid-base disturbances. Indeed, fatalities due to cholera infection are invariably due to dehydration as a result of the action of a microbial toxin interfering with normal water and electrolyte reabsorption in the colon. Additionally, the liver, one of the gastrointestinal-associated organs, contributes to acid-base balance through its role in ammonia metabolism and urea synthesis.

Changes in fluid balance, within an individual, can also occur frequently even in times of good health. Many of us suffer the side-effects, notably headache, of periods of especially underhydration, causing tissue cells to shrink, due simply to deficient fluid intake even though we are in all other respects ‘well’. Cellular overhydration resulting in cellular swelling is much more likely to be due to a homeostatic abnormality, but headache is also an early sign of fluid overload. Severe changes in hydration may occur after physical trauma resulting in injury to the body; this includes major surgery, so during the ‘post-op’ recovery period, surgeons and intensive-care physicians monitor fluid and electrolyte balance very carefully in their patients.

It is not only volume changes across cell membranes that are associated with pathology. If the volume of blood in the veins and arteries increases, the individual will suffer from high blood pressure (hypertension), a state which can have serious effects on, for example, the cardiovascular system, leading possibly to a stroke. Kidney damage can be either the cause or effect of hypertension. Conversely, if blood volume decreases (hypovolaemia) the individual will have a low blood pressure (hypotension), and all tissues are at risk as the delivery of oxygen and nutrients becomes compromised. Restoration of blood volume following a serious bleed as might occur after acute trauma, a serious accident for example, is a medical emergency requiring prompt treatment if multiple organ failure is to be avoided.

The physiological mechanisms that control fluid and electrolyte balance are subtle and often interrelated. For example, Na^+ and K^+ handling in the kidney are functionally linked with proton (H^+) secretion and the disorders that may occur in pathology are often complex. This short text is designed as an introduction to basic principles of ion and fluid homeostasis, causes and consequences of imbalance, and the role of the clinical laboratory in the assessment of such imbalances.

SECTION 1.ii

Water

Biologically, water is the most valuable resource on the planet, yet in most wealthy industrialised nations, its availability is largely taken for granted. Many organisms can exist without oxygen, but none can survive without water, as it is the solvent upon which life is dependent. Particular physical properties of water (e.g. thermal characteristics, density, viscosity and high boiling point), its chemical properties, such as its electrical polarity, and its tendency to dissociate into protons (H^+) and hydroxyl ions (OH^-), help to create the ideal chemical environment in which biochemical reactions can occur.

The polarity of water molecules illustrated in Figure 1.3 arises because the electron cloud associated with the covalent bonds between the O and H atoms is asymmetrical. The higher electronegativity of the oxygen atom pulls the electron of the hydrogen atom towards itself. This, plus the fact that each molecule is non-linear, creates definite areas of positive and negative charge which enables water molecules to interact with each other and with other compounds that are charged or ionised. Theoretically, H_2O should be a gas similar to, for example, hydrogen sulphide (H_2S) or ammonia (NH_3), but strong mutual intermolecular attraction (cohesion), arising from the polar nature of water, means that individual molecules associate so closely together that the bulk material is liquid. It is the cohesive nature of water that determines other particular characteristics, for example, freezing point temperature and volatility (tendency to evaporate and to boil). These concepts are further developed in Section 1.iii.

In a very small proportion of the water molecules present, the strain placed upon the covalent bond by its polarity is so great that the bond actually 'breaks' and the molecule ionises. The ionised species may re-associate, creating an equilibrium:

$$H_2O \rightleftharpoons H^+ + OH^- \tag{1}$$

Essential Fluid, Electrolyte and pH Homeostasis, First Edition. Gillian Cockerill and Stephen Reed.

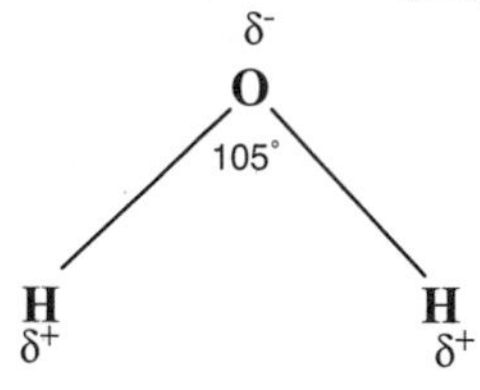

Figure 1.3 Polarity of water molecule

Strictly speaking, 'free' protons (H^+) do not exist in aqueous solution (Equation 1). Rather, they are bound with another molecule of H_2O to form H_3O^+ called a hydroxonium ion (Equation 2). For convenience however, we invariably refer to H^+.

$$H_2O + H_2O \rightleftharpoons H_3O^+ + OH^- \qquad (2)$$

The weight and size of the arrows used above indicate that the extent of ionisation is very small and most water molecules are *not* dissociated. In any equilibrium reaction such as shown above, the degree of ionisation, commonly referred to as the point of equilibrium, can be quantified thus:

$$K_{eq} = \frac{\text{Concentration of the products}}{\text{Concentration of the reactants}} \qquad (3)$$

where K_{eq} is the equilibrium constant. For the special situation with water, K_{eq} is replaced with the symbol K_w. Substituting,

$$K_w = \frac{[H^+] \times [OH^-]}{H_2O} \qquad (4)$$

where the square brackets indicate molar concentration. At equilibrium, the $[H_2O]$ is so large in comparison with $[H^+]$ and $[OH^-]$ that it is ignored, and to a good approximation,

$$K_w^{/} = [H^+] \times [OH^-] \qquad (5)$$

For *pure* water at 25°C, $K_w = 1.008 \times 10^{-14}$ which, for simplicity, is taken as 1×10^{-14}.

Equation 1 above shows us that the number of $H^+ = OH^-$, so therefore their relative concentrations must also be equal. We can state that $[H^+] = 1 \times 10^{-7}$ mol/L and $[OH^-]$ must also $= 1 \times 10^{-7}$ mol/L (i.e. $1 \times 10^{-7} \times 1 \times 10^{-7} = 1 \times 10^{-14}$).

Note that K_w is temperature-dependent and as the temperature rises, so too does K_w. For example,

at 18°C, $K_w = 0.64 \times 10^{-14}$ ($= 6.4 \times 10^{-15}$)
at 35°C, $K_w = 2.1 \times 10^{-14}$, and
at 40°C, $K_w = 2.92 \times 10^{-14}$.

A mathematically more convenient way to express the dissociation of water is as pK_w which is derived from K_w thus:

$$pK_w = -\log_{10} K_w \quad \text{(note the negative sign)}$$
$$\text{or} \quad pK_w = \log(1/K_w)$$

Again substituting, at 25°C, $pK_w = -\log_{10}(1 \times 10^{-14})$. The negative log of 10^{-14} is 14:

$$\therefore pK_w = 14$$

Using exactly the same logic and notation, we can derive a pH value:

$$pH = -\log_{10}[H^+] \qquad \text{(where } [H^+] \text{ is expressed in mol/L)}$$

if $[H^+] = 1 \times 10^{-7}$ at 25°C
then $pH = -\log_{10}(1 \times 10^{-7})$
The negative log of 10^{-7} is 7 $\therefore pH = 7$

Furthermore, because $[OH^-]$ also equals 1×10^{-7}, the pOH[3] must also be 7 and we can state that [acid] = [base]. So, at 25°C pure water is neutral at pH 7.

In addition to being the solvent in which biochemical reactions occur, water may also take part in reactions. Hydrolysis, for example, is the breaking of covalent bonds by the insertion of a molecule of water to cleave a large molecule into smaller molecules. Such reactions are often found in metabolism, and in particular in digestive processes in the gut. Hydration reactions involve the incorporation of water into a molecule creating a larger molecule.

[3]The pOH scale is not widely used.

(i) hydrolysis

$$\text{peptide} + \text{water} \longrightarrow \text{amino acids}$$

(ii) hydration

$$\text{carbon dioxide} + \text{water} \longrightarrow \text{carbonic acid}$$

$$CO_2 + H_2O \longrightarrow H_2CO_3$$

The hydration of carbon dioxide shown in (ii) is an important reaction that we will meet frequently in Part 3 of this text, which focuses on acid-base balance.

SECTION 1.iii

Solutions: concentrations and colligative properties of solutes

A solution consists of a solvent, which in body fluid systems is of course water, and one or more dissolved solutes. Body fluids are complex mixtures of solutes; for a full understanding of normal and abnormal physiological states, and of the interpretation of laboratory data, we must consider the concentration of various body fluids and how their specific chemical compositions affect their physicochemical bulk properties.

Concentration of a solution is simply the *ratio* between the amount(s) of solute(s) dissolved in a specified volume of solvent; the correct SI unit is mol/L (sometimes abbreviated to M, but *not* M/L), but typical units of concentration for physiological fluids are μmol/L (μM) or mmol/L (mM). Some laboratories outside the UK and Europe still use non-SI units such as mg/dL (also written as mg/100 mL), and occasionally hormones are measured in units such as pg/mL, so the student should be aware of such units as they are used in some textbooks.

The conversion from grams to moles, mg to mmol or μg to μmol is probably familiar to you:

$$\text{number of moles} = \frac{\text{the actual weight of the ion or molecule}}{\text{the atomic or formula weight of the ion or molecule}} \quad (6)$$

Essential Fluid, Electrolyte and pH Homeostasis, First Edition. Gillian Cockerill and Stephen Reed.

For example:

Atomic weight of Na = 23
23 mg of sodium = 1 mmol
23 µg sodium = 1 µmol
100 mmol = 2300 mg

Molecular weight of glucose ($C_6H_{12}O_6$) = 180
180 mg glucose = 1 mmol
180 µg glucose = 1 µmol
5.0 mmol = 900 mg

To express molar concentrations, it is necessary to ensure that the volume of solvent is in litres, so if a concentration is quoted as 'per dL', (i.e. per 100 mL), the number must be multiplied by 10 to bring it to 'per 1000 mL' before dividing by the atomic or formula weight.

The *ratio* will change if the proportion of *either* the solute *or* solvent changes. For example, the value for plasma sodium will rise from a normal of 140 mmol/L if either the amount of sodium present increases or if the volume of water in which the sodium is dissolved decreases; conversely, plasma sodium concentration falls if the amount of sodium falls or if the volume of plasma rises. Stated in another way, the solute concentration rises if the concentration of the solvent falls, and vice versa.

There is one other concentration unit of physiological importance that we will meet later: the mEq/L (milli-equivalents per litre). An 'equivalent' (Eq) in chemical terms represents a measure of an ion to combine with or displace hydrogen from a compound. In effect, a chemical equivalent is a function of the valency or charge on an ion. The range of equivalent values we encounter in a physiological setting is always milli-equivalent (mEq). For monovalent ions such as sodium, potassium, protons, bicarbonate and chloride, 1 mmol/L = 1 mEq/L, but for divalent ions (e.g. calcium or magnesium), 1 mmol/L = 2 mEq/L.

The presence of solutes confers certain physical properties on the solution which are different to those of the pure solvent. By definition, pure water freezes at 0°C and boils at 100°C, but water that contains dissolved and/or solvated chemical species freezes below 0°C (hence the importance of 'salting' roads during very cold weather to prevent the formation of potentially dangerous icy surfaces) and boils at a temperature above 100°C. Common experience shows us that a

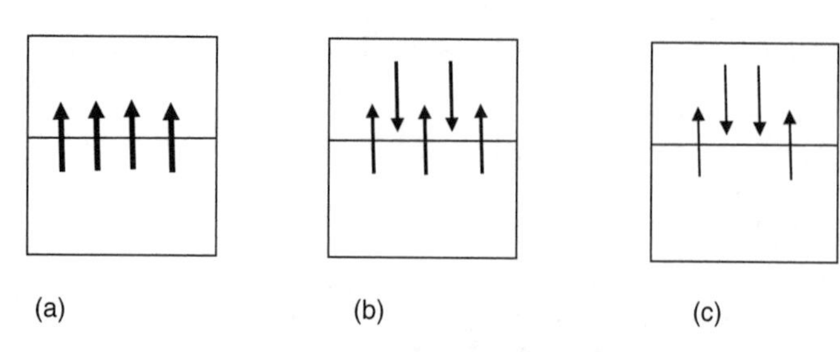

In (a) the air space is dry so liquid evaporation occurs easily but as the air becomes more moist (b) the downward pressure of the vapour in the air slows the rate of evaporation from the liquid surface. In (c) an equilibrium is reached where the tendency to evaporate is opposed by the downward force of the vapour pressure.

Figure 1.4 Vapour pressure

liquid that has an interface with a gaseous phase has a tendency to evaporate. In a closed vessel, the rate at which water molecules vaporise from the solution will initially exceed their rate of condensation from the gas phase. The pressure created by the vapour molecules will increase, creating a downward pressure on the liquid surface until the vapour (moist gas) above the liquid surface becomes fully saturated (see Figure 1.4). At this point, an equilibrium is established when the rates of evaporation and condensation are equal and the force exerted by the vapour opposing further evaporation is termed the saturated vapour pressure (SVP). Above the SVP there will be net condensation as water comes out of the vapour phase and re-forms into liquid water. We are familiar with this concept as the dew we see on the ground on cold mornings. Critically, the actual value for SVP is temperature-dependent, so the lower the ambient temperature, the lower the 'dew point' at which water vapour condenses.

Such effects due to the presence of solutes, known as colligative properties of the chemical particles present in the solvent, are a function of the number of solute particles present. The impact of these colligative properties can be described by a few 'laws'[4] of physical

[4]'Laws' are invariable demonstrable relationships observed in nature, but in practice, the mathematical relationship given above only really holds true for dilute solutions.

chemistry and can be described in general mathematical terms thus:

$$\begin{matrix}\text{Observed property} \\ \text{of a solution}\end{matrix} = \left\{\begin{matrix}\text{native property of} \\ \text{the pure solvent}\end{matrix}\right\} \times \left\{\begin{matrix}\text{mole fraction of} \\ \text{the solvent}\end{matrix}\right\} \quad (7)$$

Mole fraction is a measure of the overall ratio of solvent:solute in a solution, and is a function of the *total number* and not the *type* of dissolved ions, molecules and complexes present in the solution. Note here it is the total mass of dissolved chemical species as a ratio to the mass of solvent which determines the effect. When 1 mole of a non-volatile, non-dissociating, non-associating and non-reacting solute is added to 1 kg of pure water (a 1 *molal* solution) at atmospheric pressure, the following effects are apparent in the solution:

The freezing point is depressed by 1.858°C ($\approx$ 1.86°C)
The boiling point is elevated by 0.52°C
The vapour pressure is decreased by 40 Pa[5] (= 0.04 kPa, $\approx$ 0.3 mmHg)[6]
An osmotic pressure is exhibited (2.27 kPa $\approx$ 1700 mmHg $\approx$ 22.4 atmospheres)

Where a solute dissolves into two (e.g. NaCl) or three (e.g. Na_2HPO_4) species, the effects are increased linearly, so the change in freezing point would be $2 \times -1.86 = -3.72$°C for 1 molal NaCl, but a 1 molal solution of Na_2HPO_4 would freeze at $3 \times -1.86 = -5.58$°C. Physiological solutions are dilute so their total solute content is measured in the millimolal, rather than molal, range. As we shall see in Section 1.xiii, the unit milliosmolar (or milliosmol/L (mOsm/L)) may be encountered to emphasise the osmotic effect of solutes in solution.

[5] Pa = Pascal, the SI unit of pressure; physiological values are in kilopascals (kPa). Take care not to confuse pKa with kPa.
[6] 1 mmHg (an older unit of gas pressure) = 0.133 kPa.

Self-assessment exercise 1.1

Atomic weights: (H = 1; C =12; N = 14; O = 16, Na = 23, K = 39).

1. Calculate the pH of pure water at 18°C and 40°C.
2. From your answer to (a) above, what can you surmise about pH neutrality at body temperature?
3. Normal arterial whole blood pH falls in the range 7.35 to 7.45. Express these two values in nmol/L hydrogen ion concentration (1 nmol = 10^{-9} mol).
4. The measured osmolality of a plasma sample was found to be 295 mOsm/L. Calculate the actual depression of freezing point as measured by the analyser.
5. Study the table below.

Molecular property	CH_4	NH_3	H_2O
Molecular weight	16	17	18
Freezing point °C	−180	−80	0
Boiling point °C	−160	−35	100
Relative viscosity	0.1	0.25	1

   How can the differences revealed above be explained?
6. Given that a normal plasma [K] is approximately 4.0 mmol/L and a typical plasma glucose concentration is 5.0 mmol/L, how many mg of K and glucose are there in 100 mL of plasma?
7. Express 140 mmol/L sodium as mEq/L and as mg/L.
8. Express 1.5 mg/100 mL of creatinine ($C_4H_7N_3O$) in SI units.

Essential Fluid, Electrolyte and pH Homeostasis, First Edition. Gillian Cockerill and Stephen Reed.

9. If a sample of plasma has a calcium concentration of 2.5 mmol/L and a potassium concentration of 136.5 mg/L, what are these concentrations expressed in mEq/L?
10. A sample of urine freezes at −0.84°C. What is the molality of this sample?
11. Convert (i) 40 mmHg to kPa, and (ii) 13.0 kPa to mmHg.

Check your answers *before* continuing to Section 1.v.

SECTION 1.v

Acids and bases

Cellular metabolism involves numerous intermediates which are acids, and production of such acids constitutes a significant part of the daily acid challenge to the body. The Lowry-Bronsted explanation of acids and bases defines an acid as a proton donor and a base as a proton acceptor; both definitions assume that the reactions are occurring in an aqueous environment.

The strength of an acid or a base is a function of its 'willingness' to donate or accept a proton. Thus a strong acid donates its proton easily and a strong base accepts a proton very easily. Examples of strong acids include hydrochloric acid (HCl; $HCl \longrightarrow H^+ + Cl^-$) and sulphuric acid ($H_2SO_4$; $H_2SO_4 \longrightarrow 2H^+ + SO_4{}^{2-}$); both of these acids are fully ionised (dissociated or deprotonated).

Weak acids are poor proton donors, i.e. they do not ionise fully, so only a small fraction of the acidic molecules present will furnish a free proton, establishing an equilibrium between the bound and the ionised fractions. Metabolic acids are weak acids, including, for example, lactic acid ($CH_3.CHOH.COOH$) and carbonic acid (H_2CO_3), thus:

$$CH_3.CHOH.COOH \rightleftharpoons CH_3.CHOH.COO^- + H^+$$

and

$$H_2CO_3 \rightleftharpoons HCO_3{}^- + H^+$$

The 'two-way' arrows indicate a forward reaction (left to right, dissociation) which donates the proton, and a reverse reaction (right to left, re-association) in which a proton is accepted Note that the heavy arrow indicates that the chemical equilibrium lies to the left. It follows

Essential Fluid, Electrolyte and pH Homeostasis, First Edition. Gillian Cockerill and Stephen Reed.

from Equation 3 (Section 1.ii) that:

$$K_{eq} = \frac{[CH_3.CHOH.COO^-] \times [H^+]}{[CH_3.CHOH.COOH]}$$

Here, K_{eq} is, specifically, the weak acid dissociation constant, K_a, which is more conveniently expressed as the pKa where:

$$pK_a = -\log_{10}[K_a]$$

and thus,

$$K_a = \text{antilog}(-pK_a)$$

The pK_a is the acid dissociation constant and gives a measure of the tendency of the weak acid to donate its proton: the lower the value, the stronger the acid (i.e. greater tendency to dissociate). Typically, pK_a values fall in the range 3 to 6. For example,

lactic acid $pK_a = 3.8$ is stronger than . . .

acetic acid $pK_a = 4.7$

To be exact, the pK_a value defines the pH at which the acid group exists as the protonated (COOH) and deprotonated (COO^-) forms in equal concentration. When the actual pH of the solution containing the weak acid is greater then the pK_a of that weak acid, the chemical equilibrium moves to the right and more than 50% of the total weak acid molecules exist as anions and protons.

Some weak acids of physiological importance are 'polyprotic', that is to say, they have within their chemical structure two or even three ionisable or 'donatable' protons. A weak acid such as phosphate H_3PO_4 can donate, one at a time, three protons, and so has three pK_a values. Examples of well-known weak polyprotic metabolic acids include citric acid (triprotic) and most other compounds within the Krebs TCA cycle which are diprotic (see examples below: the ionisable protons are highlighted in bold face).

Citric acid

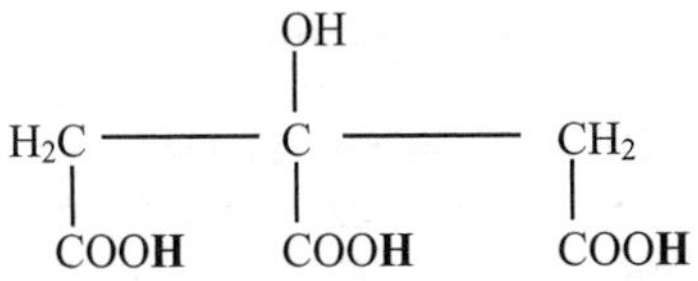

```
                 OH
                 |
H2C ——————————— C ——————————— CH2
 |               |              |
COO**H**        COO**H**       COO**H**
```

Succinic acid

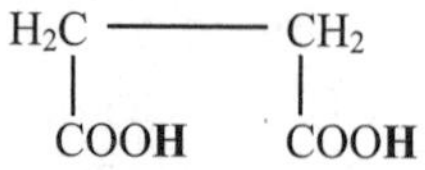

$$\begin{array}{ccc} H_2C & \text{———} & CH_2 \\ | & & | \\ COOH & & COOH \end{array}$$

Table 1.1 Dissociation constants for typical physiological weak acids

Weak acid	K_{a1}	K_{a2}	K_{a3}
Acetic acid (monoprotic)	1.83×10^{-5}	–	–
Succinic acid (diprotic)	6.6×10^{-5}	2.75×10^{-6}	–
Citric acid (triprotic)	8.1×10^{-4}	1.77×10^{-5}	3.89×10^{-6}
Phosphoric acid (triprotic)	7.5×10^{-3}	1.6×10^{-7}	4.8×10^{-13}

Note: Quoted values for K_a or pK_a are temperature-dependent.

Where a weak acid is polyprotic, the ionisation of protons is progressive, meaning that the donation of protons from particular carboxylic acid groups will occur in sequence and thus there will be a specific K_a value for each proton dissociation (Table 1.1 gives some examples). To illustrate with succinic acid:

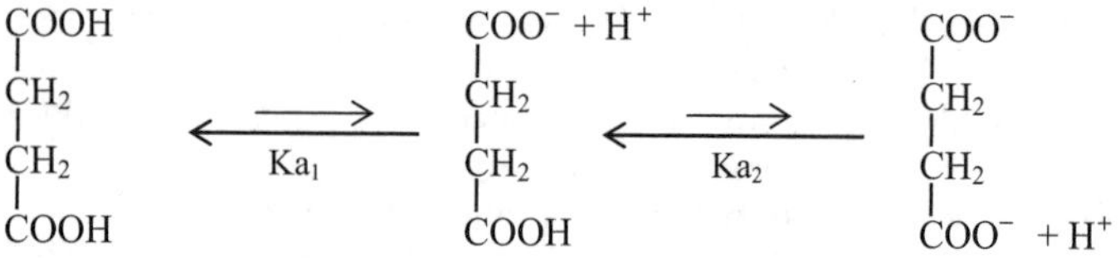

$$\begin{array}{c} COOH \\ | \\ CH_2 \\ | \\ CH_2 \\ | \\ COOH \end{array} \underset{Ka_1}{\rightleftharpoons} \begin{array}{c} COO^- + H^+ \\ | \\ CH_2 \\ | \\ CH_2 \\ | \\ COOH \end{array} \underset{Ka_2}{\rightleftharpoons} \begin{array}{c} COO^- \\ | \\ CH_2 \\ | \\ CH_2 \\ | \\ COO^- + H^+ \end{array}$$

At physiological pH (7.40 for arterial blood and 6.85 within cells), all weak metabolic acids are extensively dissociated, so we should use the name by their anion form which is the predominant form: acetic acid = acetate; succinic acid = succinate; citric acid = citrate, etc. Also of course, the 20 or so common amino acids found in proteins act as weak acids (because of their carboxylic acid group) and simultaneously act as weak bases, by virtue of their amino group. For example, the acid-base behaviour of alanine is shown in Figure 1.5.

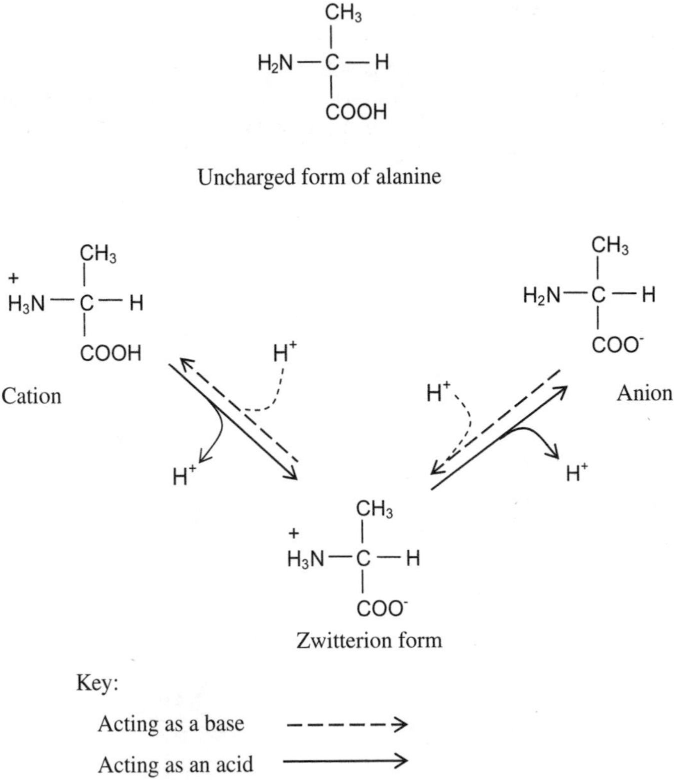

CH_3
H_2N-C-H
$COOH$

Uncharged form of alanine

$\overset{+}{H_3N}-C(CH_3)(H)-COOH$ Cation

$H_2N-C(CH_3)(H)-COO^-$ Anion

H^+ H^+ H^+ H^+

$\overset{+}{H_3N}-C(CH_3)(H)-COO^-$

Zwitterion form

Key:

Acting as a base - - - - - →

Acting as an acid ————→

Figure 1.5 Acid-base behaviour of alanine.

In practice, the uncharged alanine species does not exist in great concentration, therefore pKa_1 defines the pH at which 50% dissociation of the COOH group has occurred (zwitterion and cation forms), whilst pKa_2 is the pH where there are equal proportions of zwitterion and anion. When neutral amino acids such as alanine are combined together in proteins, the carboxylic acid and amino groups of adjacent molecules are used in peptide bond formation, so only the N-terminal and C-terminal residues show the acid-base behaviour illustrated above. However, a few amino-acid residues such as glutamate, aspartate (both dicarboxylic), histidine, arginine and lysine (which have additional proton-accepting basic groups) are able to donate or accept protons even when found within the peptide chain of a protein. Histidine is particularly important because it is the only amino acid that has a proton which is able to dissociate in the neutral pH range. The ability to donate or accept protons makes proteins very important buffers, as we shall see in Part 3.

SECTION 1.vi

Buffers and the Henderson-Hasselbalch equation

The normal processes of metabolism produce large quantities of acid each day, which if left unchecked would very soon cause irreparable damage to cells and the organism. Efficient means of dealing with the acid insult, including chemical buffering, are therefore of the utmost importance for survival.

A buffer resists changes in pH of a solution and is composed of a weak acid (shown here as HB) and a salt (represented as NaB) of that weak acid.

e.g.

$$NaB \longrightarrow Na^+ + B^- \quad \text{a salt is fully dissociated}$$

$$HB \rightleftharpoons H^+ + B^- \quad \text{a weak acid is only weakly dissociated}$$

The salt provides anions (B^-) to react with, and therefore neutralise, any addition of protons. The weak acid is present to furnish (by dissociation) more anions as they are 'used up' by added protons. A weak acid obeys the laws of chemical equilibria, so when some anions are 'removed' (as a result of the neutralisation), the position of the equilibrium of the weak acid will change to liberate more anions. The addition of what is in practice a relatively small number of anions has a negligible effect, as the total concentration of B^- supplied by the salt and the acid dissociation together is far greater than any reasonable increase.

Essential Fluid, Electrolyte and pH Homeostasis, First Edition. Gillian Cockerill and Stephen Reed.

For example, the weak acid could be acetic (ethanoic) acid and the salt would be sodium acetate.

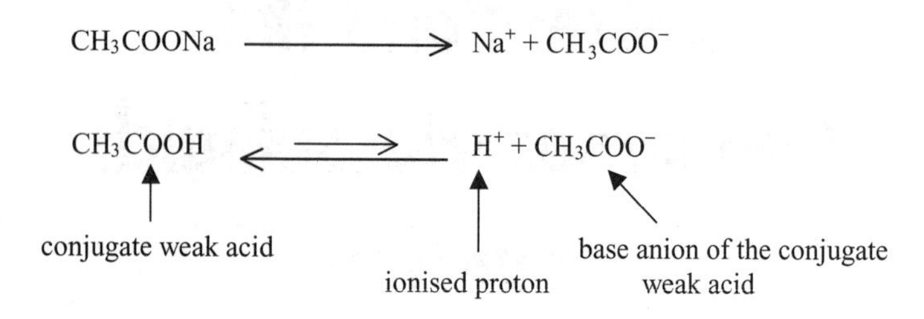

Note that the difference between the conjugate acid and the base is only a single proton (i.e. the base anion is a 'de-protonated' form of the weak acid). Note also that some weak acids have two or more ionisable protons and thus have two or three conjugate pairs of protonated (acid) and deprotonated (base anion) forms.

Recall from Section 1.v that the acid dissociation constant (pKa) of a weak acid determines exactly the extent of dissociation. The relative concentrations of the components of a buffer system at a given pH are defined by the Henderson-Hasselbalch equation:

$$\mathrm{pH} = \mathrm{pKa} + \log_{10}\left\{\frac{[\text{base}]}{[\text{conj. weak acid}]}\right\} \tag{8}$$

Note that face brackets are used here to emphasise that we need the $\log_{10}$ of the base:acid ratio, not log of [base] divided by log of [acid].

An example calculation:

A buffer solution consists of 1.35 g/L lactic acid and 2.24 g/L sodium lactate.
If the pKa of lactic acid is 3.8, calculate the pH of the buffer solution.
Atomic weights: H = 1; C = 12; O = 16; Na = 23.

Firstly, convert g/L to molar concentrations (see Equation 6 in Section 1.iii):

Molecular formula for lactic acid = $CH_3.CHOH.COOH$
Molecular weight for lactic acid = 90
Molecular formula for sodium lactate = $CH_3.CHOH.COONa$

Molecular weight for sodium lactate = 112

$$[\text{weak acid}] = \frac{1.35}{90} = 0.015\,\text{mol/L}(= 15\,\text{mmol/L})$$

$$[\text{base}] = \frac{2.24}{112} = 0.02\,\text{mol/L}(= 20\,\text{mmol/L})$$

$$\text{pH} = \text{pKa} + \log_{10}[\text{base}]\left\{\frac{[\text{base}]}{[\text{conj. weak acid}]}\right\}$$

base:acid ratio = 0.02/0.015 = 1.333
(*Note*: [base] and [conj. acid] must be in same units)

$$\log_{10} 1.3333 = 0.125$$

$$\therefore \text{pH} = 3.8 + 0.125 = \mathbf{3.925}\,(3.93 \text{ to 3 significant figures})$$

We can predict from the Henderson-Hasselbalch equation that buffers are most efficient in 'mopping-up' free protons when there are approximately *equal* concentrations of base and acid present, i.e. acid:base ratio of 1:1 and pH = pK_a. Indeed, in a test-tube, buffer efficiency is maximal when the actual solution pH = $pK_a \pm 1$ pH unit. This rule does not always hold true for physiological systems, however, because unlike the situation in a test-tube, homeostatic regulatory mechanisms can 'top-up' the concentration of the individual conjugate pairs, perhaps in response to a pH challenge to ensure efficiency of the buffer system.

SECTION 1.vii

Self-assessment exercise 1.2

1. Complete the following table by calculating values indicated by ??
   Arrange the acids as a list with the strongest at the top and weakest at the bottom.

Acid	K_a	pK_a
Acetic	1.83×10^{-5}	??
Carbonic	7.9×10^{-7}	??
Acetoacetic	2.6×10^{-4}	??
Fumaric	$Ka_1 = 9.3 \times 10^{-4}$	??
	$Ka_2 = 2.88 \times 10^{-5}$	??
Phosphoric	$Ka_2 = 1.6 \times 10^{-7}$	??
Succinic	$Ka_1 = 6.6 \times 10^{-5}$	??
	$Ka_2 = 2.75 \times 10^{-6}$	??
Pyruvic	??	2.50
Aspartic side chain COOH	??	$pKa_2 = 3.9$
Histidine	??	$pKa_3 = 6.0$

2. How does knowledge of the ionic behaviour of amino acids help us understand the buffer action of proteins?
3. Write balanced chemical equations to show the dissociation of pyruvic acid ($CH_3.CO.COOH$) and of phosphoric acid (H_3PO_4).
4. State, giving reasons, whether (i) histidine (imidazole side chain $pK_a = 6.0$), and (ii) the side chain amino group of arginine ($pK_a = 12.6$) would be protonated or deprotonated at typical cytosolic pH of 6.85.

5. What proportions (ratio) of lactic acid and sodium lactate would be required to prepare a buffer with a final pH of 3.5?
6. Use your answer to question 5 above and state how many grams of lactic acid and sodium lactate would be needed to prepare 2.5 litres of the buffer (pH = 3.5).
7. Refer to the diagram of citric acid shown in Section 1.v above. Draw a chemical equation to show the successive dissociation of the three ionisable protons. (*Note*: The COOH attached to the central carbon has the highest pK_a.)
8. Figure 1.5 shows the ionic behaviour of alanine where the side chain R group is $–CH_3$. Given that the side chain for lysine is $–CH_2–CH_2–CH_2–CH_2–NH_2$, draw chemical structures showing the different ionic forms.
9. Complete the following table by finding values where '??' is shown.

	Conjugate weak acid concentration (mol/L)	Base concentration (mol/L)	pK_a of conjugate weak acid	Final pH of the buffer solution
i	0.05	0.05	4.8	??
ii	0.02	0.16	3.5	??
iii	0.24	0.03	3.5	??
iv	25×10^{-3}	1.2×10^{-2}	5.1	??
v	1×10^{-3}	2×10^{-3}	??	6.8

10. What would be the base:acid ratio in a buffer if the final pH of the solution is 7.40 when the pK_a is 6.1? Would you expect this buffer to be effective? Give reasons.
11. Given that pKa_2 for phosphoric acid ($H_2PO_4^- \rightleftharpoons HPO_4^{2-} + H^+$) is 6.8, calculate the base:acid ratio at normal blood pH of 7.40 (actual concentration values for $H_2PO_4^-$ and HPO_4^{2-} are not required).

Check your answers *before* proceeding to Section 1.viii.

SECTION 1.viii

Body fluids and their composition: overview

Assuming that 1 litre of water weighs 1 kg, we can say that approximately 60% of the body weight of an average 70 kg man (55% of body weight in females) is water. Total body water is divided between two main 'compartments': extracellular fluid (ECF), which comprises about a third of the total body water volume; and the remaining two-thirds is in the intracellular fluid (ICF) compartment. Further, the ECF is composed of interstitial fluid (ISF), which is fluid between and around tissue cells, blood plasma, and transcellular fluid, which is that component of the ECF that has crossed an epithelial barrier, i.e. has been secreted. Such transcellular fluids include cerebrospinal fluid which bathes and supports the central nervous system (see later in this Section), vitreous humour in the eyes, synovial fluid found in joints of the skeleton and the digestive fluids secreted into the gastrointestinal tract. This distribution of fluid is represented in Figure 1.6. It should be noted that fluid content of the body decreases with age; for example, approximately 75% of the body weight of an infant is water, whilst in the elderly (70+ years) water accounts for only 45% of body weight (see Table 1.2).

Body fluids are complex mixtures of large (e.g. protein molecules, lipoprotein particles) and small (metabolites and ions) solutes. Relatively small polar or charged compounds are easily dissolved in water, whilst larger charged molecules such as proteins and lipoprotein complexes become solvated, i.e. suspended as a fairly viscous colloid of finely dispersed particles. Cellular cytosol and plasma are both protein-rich and are examples of colloids.

The concentrations of principal ions and representative other solutes, in the body are shown in Table 1.3.

Essential Fluid, Electrolyte and pH Homeostasis, First Edition. Gillian Cockerill and Stephen Reed.

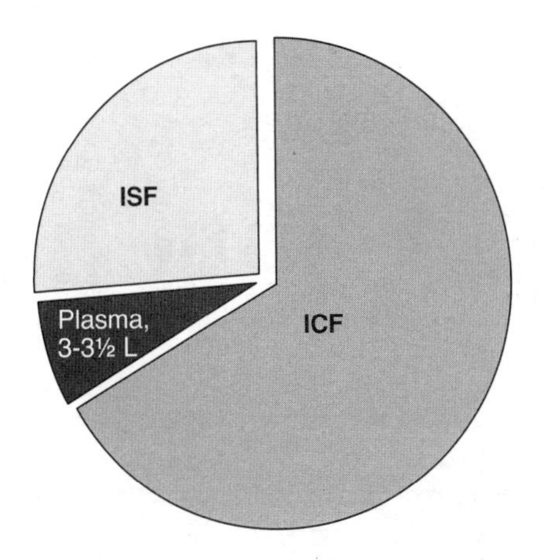

ISF, interstitial fluid; ICF, intracellular fluid.
ISF and plasma together constitute the extracellular volume.

Figure 1.6 Total body water

Table 1.2 Total body water: effects of age and body shape

Time of life	Water content of body by weight, average build (%)	Water content of body by weight, thin build (%)	Water content of body by weight, fat build (%)
Infants	75	80	65
Young males (<30 yrs)	60	65	55
Young females	55	60	45
Elderly males (>65 yrs)	50	55	45
Elderly females	45	50	40

Note: Total body fluid volume is controlled very sensitively so that typical day-to-day fluctuation is no more than 0.2%.

ECF

The major part of the ECF is plasma, a slightly viscous fluid which is, by weight, approximately 95% water with the remaining 5% of solute being, by weight, mostly protein. In health the chemical composition of plasma varies only slightly due to normal physiological variation, possibly following some predetermined biorhythm, be it within the

Table 1.3 Major solutes of body fluids

	ECF		
	Plasma	ISF	ICF
Cations			
Sodium, Na	135–145 mmol/L	140–145 mmol/L	10–15 mmol/L
Potassium, K	3.6–4.5 mmol/L	4.0 mmol/L	145–150 mmol/L
Calcium, Ca	2.20–2.55 mmol/L (total calcium)	1.25 mmol/L (as free ionised Ca^{2+})	<1 mmol/L (free ionised Ca^{2+})
	1.10–1.25 mmol/L (ionised calcium)		
Magnesium, Mg	0.7–1.1 mmol/L	0.5 mmol/L	2.5 mmol/L
Anions			
Bicarbonate, HCO_3^-	22–28 mmol/L	28–32 mmol/L	10–12 mmol/L
Chloride, Cl	95–105 mmol/L	115–120 mmol/L	2–5 mmol/L
Phosphate[a]	0.8–1.3 mmol/L	0.8–1.3 mmol/L	75 mmol/L
Sulphate, SO_4^{2-}	~0.8 mmol/L	~0.8 mmol/L	5 mmol/L
Organic acids[b]	<5 mmol/L	<5 mmol/L	5 mmol/L
Protein[c]	55–75 g/L	Difficult to quantify; 'trace'	180 g/L
	10–15 mEq/L		35–42 mEq/L
Others			
Glucose	3.0–5.0 mmol/L	3.0–5.0 mmol/L	<0.5 mmol/L
Urea	3.0–7.0 mmol/L	3.0–7.0 mmol/L	1.0–2.0 mmol/L[d]

[a] Phosphate can exist as PO_4^{3-}, HPO_4^{2-} or $H_2PO_4^-$, but it is the HPO_4^{2-} form which predominates at normal blood pH. The $H_2PO_4^-$ form is found mainly in urine where typical pH is 5.5–6.5.

[b] The organic anions in ECF represent a disparate group of small molecules, but it is the overall charge of ~7–8 mEq/L carried by these molecules which is important.

[c] Proteinate. The charge carried by proteins is pH-dependent. At normal blood pH the average charge carried on protein molecules is approximately 12^-.

[d] Urea is synthesised in the liver so the local concentration may be higher than the value quoted. Because urea is diffusible across cell membranes, there may be some in non-hepatic cells.

course of a day (diurnal rhythm) or monthly or even seasonal. The usefulness of plasma as an analytical matrix is due to the fact that metabolic changes and physiological dysfunction occurring within tissue cells often reflect with changes in the composition of plasma. Also, taking blood is usually a simple and relatively pain-free process in most cases.

Cerebrospinal fluid (CSF) is normally a clear colourless fluid that supports and protects the central nervous system. The brain of an adult weighs around 1.5 kg, but the buoyancy effect of the CSF reduces the effective weight of the organ to less than a third of that value. The total volume of CSF at any one moment is approximately 140 mL, yet the rate of production of CSF is just over 700 mL per day, indicating that resorption of CSF must also occur at the same rate if the constant volume is to be maintained.

CSF arises from three sources:

(1) ultrafiltration of plasma perfusing the choroid plexi (singular = plexus), which are two structures located within the lateral ventricles (cavities) of the brain. This process accounts for approximately 70% of total CSF production;
(2) secretion directly from the blood cerebral capillary system (20% of the total); and
(3) metabolic water (10%) derived from the oxidation of glucose which is the major and preferred fuel of the central nervous system.

Once produced, CSF circulates in the sub-arachnoid space between the brain tissue and the skull and also along the length of the spinal canal surrounding the cord.

Unlike taking blood, the collection of CSF is neither simple nor convenient for the patient. A needle is inserted into the lumbar (lowest) region of the spine, below the level at which the spinal cord ends. An alternative to the lumbar puncture is collection of CSF from the base of the skull by insertion of a needle into the back of the neck; this process is not without risk.

Chemical analysis of CSF may however be of diagnostic significance. A comparison of the chemical compositions of plasma and CSF is shown in Table 1.4. Chemical analysis of CSF for glucose and/or protein concentrations may be diagnostically useful as an adjunct to microbiological investigation of meningitis, or for the detection of pathologies such as malignancy or neurodegenerative disease. The

Table 1.4 A comparison of the chemical compositions of CSF and plasma

Analyte	CSF*	Plasma
Water	99%	94%
Glucose	3.0 mmol/L	5.0 mmol/L
Total protein	0.05 g/L (50 mg/L)	70 g/L
Sodium	140 mmol/L	140 mmol/L
Potassium	3.0 mmol/L	4.0 mmol/L
Chloride	120 mmol/L	100 mmol/L
Bicarbonate	25 mmol/L	25 mmol/L
Calcium	1.5 mmol/L	2.5 mmol/L
pH	7.34	7.40

* *Note*: There are some regional differences in chemical composition of the CSF bathing different parts of the CNS.

laboratory may be asked to identify the presence of blood or bilirubin, the yellow pigment of haem catabolism, in CSF to confirm that a sub-arachnoid haemorrhage (SAH) has occurred.

One other body fluid whose analysis has found some clinical use for diagnostic purposes is saliva. Like CSF, saliva is an ultrafiltrate of plasma and is produced at a comparable rate to CSF, typically 0.5 mL per minute, although this value varies widely due to a number of psychological and physiological factors. The attraction of saliva as a clinical sample is the ease and comfort of its collection; no needles are required, although strict adherence to a standardised protocol is necessary if meaningful results are to be obtained. The range of clinical analytes which can usefully be quantified in saliva is limited, but notable examples include 'free' (i.e. non-protein bound) steroid hormone concentration (such as cortisol or progesterone) and immunoglobulins.

The chemical natures of interstitial fluid and lymph are substantially the same as ECF except with respect to their lower protein concentrations. Lymph drainage from the gut may appear 'milky' due to the presence of fat-rich chylomicrons particles, which are formed in the enterocytes after a meal and exported not directly to the blood stream, but to the lymph.

Pleural, peritoneal and synovial fluids are examples of transcellular fluids that originate in the ECF by a process of selective filtration, so the term transudate is often used to highlight their formation. The pleural membranes line the inner wall of the chest; the peritoneum is the lining of the abdomen and the synovium provides lubrication to joints. Normally, the volumes of pleural and peritoneal fluids are small, but effusion resulting in volume expansion within the space may occur in certain pathologies such as infection or cancer. The fluid, termed an exudate, which accumulates under these conditions, has a different chemical composition to the normal and its analysis may be of diagnostic value. An abnormally large volume of peritoneal fluid which accumulates in pathology is often referred to as ascitic fluid, and the two terms are in practice interchangeable. In contrast, the chemical analysis of synovial fluid is of little use, although there is an emerging interest in the estimation of inflammatory marker proteins and trace metals such as cobalt or chromium to indicate wear of metal joint implants.

ICF

The exact water content of tissues varies slightly, and typical values given often relate to muscle and total body water, and varies inversely with the amount of adipose (fat) tissue in an individual; obese people have less total body water than thinner people of comparable total body weight (see Table 1.2).

ISF

As can be seen from the data in Table 1.3, ISF is essentially an ultra-filtrate of plasma, the most noticeable difference being in the protein concentration.

As will be discussed more fully in the next Section, there is continual transfer of fluids and the solutes they contain between compartments and parts of the body.

SECTION 1.ix

Fluid balance: (a) between fluid compartments and (b) intake and loss

(a) Fluid moves between body compartments

There is a continual exchange, a dynamic equilibrium, of water and solutes between the main fluid compartments to ensure delivery of nutrients to, and removal of waste from, all tissue cells (Figure 1.7). Additionally, large volumes of fluid are secreted each day. For example, 6–7 litres of fluid enter the gut during a 24-hour period, and the kidneys filter approximately 140 litres of plasma water each day, so clearly substantially more fluid is turned-over each day than is present in the extracellular fluid. Most of the secreted and filtered fluid is subsequently reabsorbed back into the circulatory system. Thus an 'internal balance' is established as compartments exchange fluid. The mechanisms determining fluid distribution between compartments are described more fully in Section 1.xii, page 50. The presence in abnormally high concentrations of osmotically active compounds such as ethanol can seriously upset the distribution of water between compartments. The headache which often follows excessive intake of alcoholic drinks is due to cellular dehydration as water is pulled out of tissues, notably the brain. Moreover, alcohol is a diuretic, so not only is water lost from the ICF compartment, it is also lost from the body as urine, as part of external fluid balance.

Essential Fluid, Electrolyte and pH Homeostasis, First Edition. Gillian Cockerill and Stephen Reed.

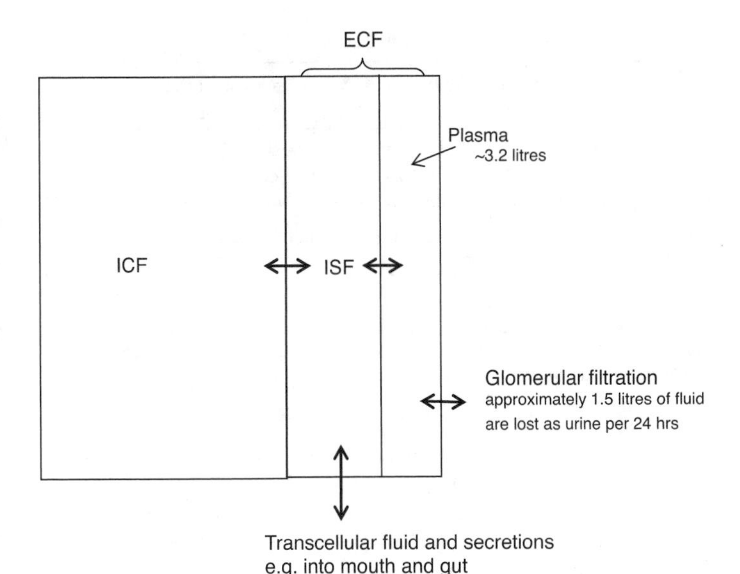

Figure 1.7 Body fluids are in dynamic 'flux' moving between compartments

(b) Intake and loss

In addition to fluid movements between compartments, we must also consider the fluid that enters and leaves the total body water volume each day, since this too contributes to the dynamic state of hydration. This may be called 'external balance'.

Note that obligatory loss is usually matched by fluid in food and that derived from metabolism. The term 'metabolic water' denotes the turnover of water as part of day-to-day catabolic and anabolic processes. For example, water is obtained from the oxidation of fuel:

$$\text{glucose} + \text{oxygen} \longrightarrow \text{carbon dioxide} + \text{water}$$
$$C_6H_{12}O_6 + 6O_2 \longrightarrow 6CO_2 + 6H_2O$$

also:

$$\text{fatty acid} + \text{oxygen} \longrightarrow \text{carbon dioxide} + \text{water}$$
$$C_{18}H_{32}O_2 + 23O_2 \longrightarrow 16CO_2 + 16H_2O$$

Table 1.5 Typical fluid balance of a healthy individual

Water added to the body fluid pool each day		Water lost from the body fluid pool each day	
As fluid intake	1500 mL[a]	As urine	1500 mL[b]
In moist food	750 mL	In faeces	100 mL
From metabolic processes	250 mL	Insensible losses[c]	900 mL
Total	2500 mL	Total	2500 mL

[a]Variable according to personal habit.
[b]Variable according to fluid intake.
[c]Insensible losses include moisture on the breath and perspiration. Fluid loss by these routes plus that lost via the gut is sometimes referred to as obligatory loss, as it is essentially uncontrollable by homeostatic means.

This gain of water is offset by its use in, for example, (i) hydrolytic and (ii) hydration reactions as described in Section 1.ii.

In the course of a day, large volumes of fluids and the solutes which they contain are 'turned over' within the body (Table 1.6 below). Daily net losses of fluids and electrolytes must be replaced by dietary intake.

Table 1.6 Typical volumes and ionic composition of some body fluids and secretions

Fluid	24-hour volume in litres	Sodium mmol/L	Potassium mmol/L	Bicarbonate mmol/L	Chloride mmol/L
Saliva	0.1	60	15	15	20
Gastric juice	2.5	70	10	<10	140
Intestinal fluids					
total	4.5	–	–	–	–
jejunal	–	135	5	10	135
ileal	–	140	5	30	125
pancreatic	1.25	130	8	85	55
bile	0.6	145	5	30	100
Sweat	0.5	40–55	8–10	0	35–45
Urine	1.5	80–120	30–70	–	70–110
Faeces	0.1	80	150	<5	<10
diarrhoea fluid	Variable	40–120	30–70	30–70	<10
colostomy	0.25	60	10–20	<5	30–45
ileostomy	0.5	50	4	<5	25

The physiological purpose of urine is to carry away from the body the water-soluble waste materials produced by metabolism each day. In health, the daily volume of urine reflects fluid intake, rather than intake having to match some presumed 'normal' urine output of around 1.5 litres. The *minimum* volume of urine needed to excrete the waste solutes without risk of them precipitating in the urinary system is about 600 mL, a value which depends to some extent on diet.

Daily fluid intake must equal the volume lost as urine. Therefore, to ensure excretion of solute waste without risk, everyone should drink at least 600 mL of fluid a day – *not* 1600–2000 mL as is widely believed. An often quoted cliché is '8 × 8', which means that to maintain 'good health', '8 fluid ounces of water must be taken 8 times a day'. A fluid ounce is approximately 240 mL so '8 × 8' equates to an intake of around 1900 mL water per 24 hours. The origin of this statement is uncertain and its physiological validity is at least questionable, but the fact that the recommendation has gone largely unchallenged has led to it becoming part of popular health dictum: a case of an accepted 'fact' arising from its frequent repetition rather than from any rigorous scientific evidence. For a more detailed critique of the '8 × 8' rule, the interested reader is referred to a review by Valtin.[7]

Furthermore, the 'form' (pure water, beverages, soft drinks, etc.) that the fluid takes is of little importance. Some people believe that only pure water 'counts' as part of the daily fluid intake, and that fluids containing caffeine are not significant contributors to fluid balance. Although many beverages contain caffeine, which is a diuretic, its concentration in tea or coffee is fairly small, typically less than 60 mg in a typical 200 mL cup of coffee, and below the amount required to achieve any significant effect on adenosine receptors of the nephron, so the impact of this chemical on water excretion is physiologically mild or even insignificant. Indeed, the extra volume of fluid lost as result of caffeine action in the kidney is less than the volume of the drink which contained the caffeine, therefore consumption of tea or coffee does lead to a small net gain of liquid. Moreover, a degree of pharmacological tolerance develops towards caffeine, so regular tea or coffee drinkers become less susceptible to its effects. Alcohol (ethanol), on the other hand, is a diuretic and

[7]Valtin, H. (2002) Am. J. Physiol. Regul. Integr. Comp. Physiol. **283**: R993–R1004.

excessive intake can lead to systemic dehydration, often apparent as a headache.

In practice, fluid intake is determined as much by personal and cultural habits as it is by physiological need. For example, every day events such chatting to friends or colleagues over a cup of coffee means that we usually drink more than the minimum volume required to ensure satisfactory excretion of waste solute. For an adult leading a sedentary lifestyle in a temperate climate, a minimum intake of around 1000 mL per day is probably wise and a more realistic intake than the 1900 mL of the '8 × 8' recommendation. Water intakes approaching or even exceeding 1500 mL per day may, however, be necessary in certain circumstances, for example in times of excessive sweating such as during periods of heavy physical work, vigorous exercise or fever, or during time spent in tropical climes. Perspiration is relatively salt-free (see Table 1.6), but in all such cases an adequate co-intake of salt is also required, hence the availability of commercial 'isotonic fluid' replacements. Also, increased fluid intake is recommended in certain pathological conditions such as for individuals who have a tendency, possibly genetically determined, to renal stone formation, where it is vital to ensure the excretion of dilute urine.

SECTION 1.x

Ionic composition and electrical neutrality

The chemical composition of body fluids varies significantly in terms of both electrolytes and non-ionic species. In order to maintain electrical neutrality, the total number of positive *charges* must equal the total number of negative *charges*. 'Equal *charges*' does not equate with 'equal *numbers*' of ions, as some may be di- or even trivalent. The electrical charge is usually expressed as mEq/L (milliequivalents per litre). For monovalent ions such as sodium, 1 mmol/L = 1 mEq/L, but for divalent ions such as calcium, 1 mmol/L = 2 mEq/L. Figure 1.8 shows the major ions in plasma and selected transcellular fluids.

It is the total number of charges on all of the anions and cations in the body fluids which ensure electrical neutrality. Expressing this in simple mathematical terms:

$$\Sigma\,(\text{cation concentration}) = \Sigma(\text{anion concentration})$$

where Σ (sigma) stands for 'the sum of . . .'

$$\text{put more simply,}\quad \Sigma C^{+} = \Sigma A^{-}$$
$$\text{and for neutrality,}\quad \Sigma C^{+} - \Sigma A^{-} = 0$$

The maintenance of equal charge distribution, expressed in mEq/L, across cell membranes is a normal physiological requirement, except when deliberate and intentional depolarisation of nerve or muscle cell membranes is needed to generate a 'signal'. To satisfy the Gibbs-Donnan equilibrium (see page 46), we must maintain equal cation and anion concentrations *within* each of the ICF and ECF compartments and *between* these two compartments. Thus if we imagine a decrease in, say, plasma bicarbonate concentration, there must be either a corresponding decrease in cation concentration or an increase in the concentration of another anion, most likely chloride rather

Essential Fluid, Electrolyte and pH Homeostasis, First Edition. Gillian Cockerill and Stephen Reed.

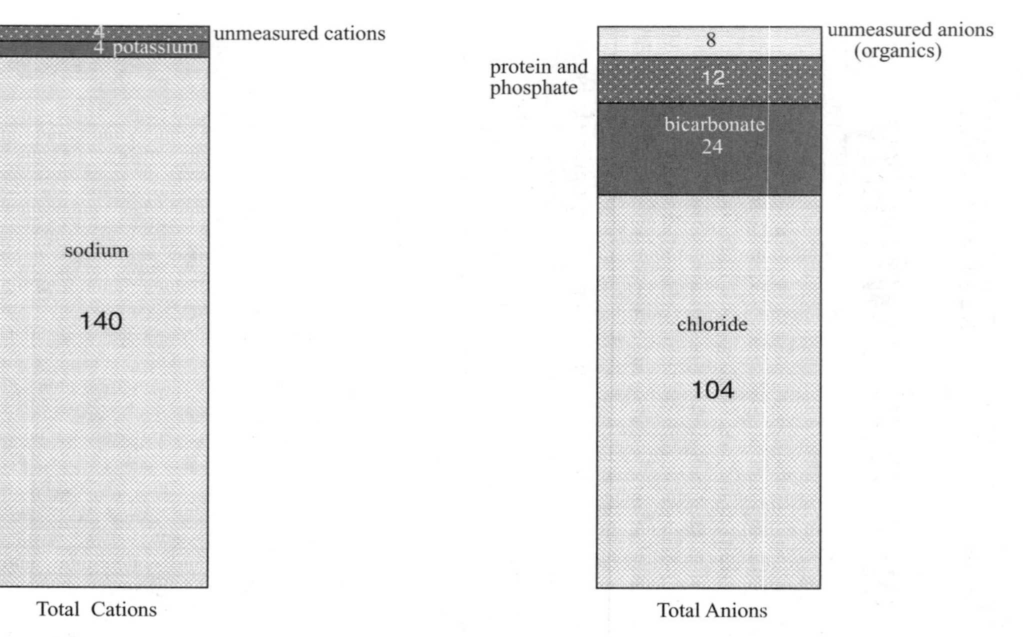

Figure 1.8 Ionogram: relative concentrations (in mEq/L) of main ions in plasma (ECF) Total cation *charge* = total anion *charge* ~148 mEq/L

than protein or phosphate. Conversely, a rise in plasma chloride concentration (hyperchloraemia) will lead to a reduction in bicarbonate concentration; this situation is often encountered in certain cases of acidaemia. It could be argued that a fall in plasma bicarbonate will be matched by a corresponding decrease in *plasma* sodium concentration, in order to maintain cation/anion balance but this is unlikely because a fall in plasma sodium would require a compensatory fall in the cation concentration in the *intracellular* compartment. In short, a fall in plasma bicarbonate will result in a rise in plasma chloride concentration thus maintaining the same total anion concentration but by different values for chloride and bicarbonate. Similar logic could apply to cation concentration in plasma; for example, a rise in $[K^+]$ would theoretically cause a fall in $[Na^+]$, but the concentration of sodium in plasma: (a) has a relatively wide physiological range; and (b) is so much greater than that of potassium that small compensatory fluctuations in $[Na^+]$ would not be immediately apparent.

Urine also is electrically neutral. If for some reason there is the need to excrete an unusually high anion load, e.g. oxoacids during a diabetic crisis, or an anionic toxin such as salicylate, there would be an equivalent loss of cation, probably sodium in the urine, possible severe enough to cause a noticeable reduction in plasma sodium concentration (hyponatraemia).

Routinely, clinical laboratories measure the plasma concentrations of sodium, potassium, bicarbonate and occasionally chloride, but as we have seen in Table 1.3, there are many other physiologically important (organic) ions present in body fluids. These 'other' organic ions all contribute to ΣC^+ and ΣA^-, but they are unmeasured by routine laboratory methods.

So now we can say

$$\Sigma C^+ = \{[\text{measured cations}]\} + \{[\text{unmeasured cations}]\}, \quad \text{and}$$

$$\Sigma A^- = \{[\text{measured anions}]\} + \{[\text{unmeasured anions}]\}$$

where [ ] signifies molar concentration. Clearly, for electrical neutrality to exist:

$$\{[\text{measured cations}]\} + \{[\text{unmeasured cations}]\} = \{[\text{measured anions}]\} + \{[\text{unmeasured anions}]\}$$

Figure 1.8 shows that quantitatively the unmeasured cation fraction is fairly small in comparison with Na^+ and K^+, which taken together constitute measured ΣC^+. Furthermore, changes in the concentrations of either calcium or magnesium are very small in comparison with Na^+ or K^+ and can be ignored. However, changes in the concentration of unmeasured anions can be physiologically significant. Thus, to a good approximation:

$$[\text{measured cations}] \approx \{[\text{measured anions}]\} + \{[\text{unmeasured anions}]\}$$

and thus

$$\{[\text{measured cations}]\} - \{[\text{measured anions}]\} = [\text{unmeasured anions}]$$

A useful derived parameter, the ANION GAP (AG), is therefore obtained:

$$\begin{aligned} AG &= ([Na^+] + [K^+]) - ([HCO_3^-] + [Cl^-]) \\ &= \sim[\text{unmeasured anions}] \\ &= (140 + 4) - (25 + 100) \end{aligned} \tag{9}$$

typical range of values for AG = 15–20 mmol/L.

A simplified version of this equation omits the [K]:

$$AG = ([Na^+]) - ([HCO_3{}^-] + [Cl^-]) = \sim[\text{unmeasured anions}]$$

Substituting typical plasma values:

$$\begin{aligned} AG &= (140) - (25 + 102) = 13\,\text{mmol/L} \\ \text{range} &= 11\text{–}16\,\text{mmol/L} \end{aligned}$$

The anion gap should not be confused with the osmolal gap (see page 58), which is a measure of the difference between the osmolality and osmolarity results and so reflects the total solute load rather than just the major ion composition. Equation (9) above suggests that the total cation concentration is higher than the total anion concentration but, and as is the case with the osmolal gap, it is important to realise that *physiologically* there is no such thing as the anion gap (because electrical neutrality is maintained); it is important to remember that the range quoted (11–16 mmol/L or 15–20 mmol/L) is actually an estimation of anions (such as proteinate, phosphate, sulphate and organic anions, e.g. urate and lactate) in plasma which are not routinely measured as part of the laboratory's repertoire of tests.

SECTION 1.xi

Water and ion distribution between compartments 1: Physical chemistry

Physiological fluids are continually being exchanged between the intracellular and extracellular spaces of the body. These processes of exchange are driven by physicochemical forces, namely (1) the requirement for osmotic balance and (2) the need to maintain electrical neutrality as described by the Gibbs-Donnan membrane equilibrium.

(a) Osmosis

Osmosis is a particular type of passive diffusion in which water molecules move across a membrane separating two compartments. The nature of the membrane and the composition of the solutions on either side of that membrane determine the rate and extent of the transfer of water between the compartments. If the membrane is freely permeable to solvent and all solutes present, there will be diffusion in both directions and eventually the two solutions will become homogeneous.

The term 'selectively permeable'[8] is usually used to describe a membrane that restricts the passage of some types of molecule but not

[8] Use of the older term *semi-permeable* is not recommended.

Essential Fluid, Electrolyte and pH Homeostasis, First Edition. Gillian Cockerill and Stephen Reed.

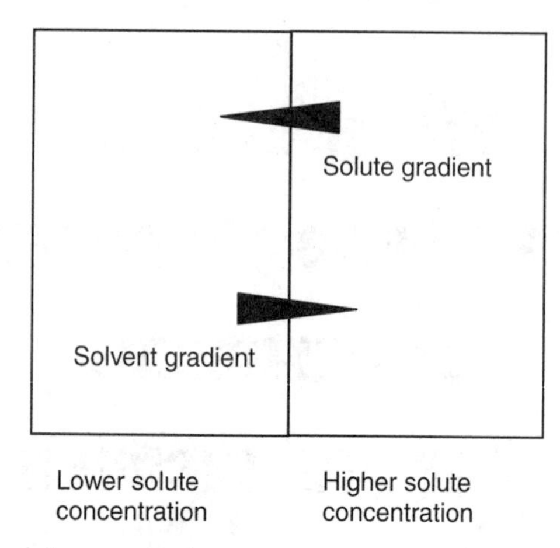

Figure 1.9 Solvent and solute gradients across a selectively permeable membrane

others. If such a membrane separates pure solvent (water) from a solution, or if it separates two solutions of different concentrations, a solute concentration gradient is established across the membrane as shown in Figure 1.9. The relative proportions of solvent and solute may be termed the mole fraction of solute.

Assuming the total volumes in the two compartments are identical, the true water volume will be lower in the solution with the higher solute concentration, as the solute particles occupy space, effectively displacing water molecules. This situation is energetically 'unstable', and so to dissipate the gradients there will be a tendency for diffusion to occur until the chemical composition (not the volumes) of the two liquids become equivalent. Assuming also that the membrane is selectively permeable to water molecules only, water will move into the compartment of higher solute concentration and thus the volumes of the two compartments will change until the solute concentrations are equal. We may explain the process occurring in slightly different ways:

- Water passes *into* the compartment with the higher solute concentration;
- There is a tendency for the movement of water in both directions across the membrane, but free flow of water *between* the compartments is impeded by the presence of solutes, so the overall rate of

diffusion between the two spaces is unequal. Water does not leave the compartment with the higher solute concentration as easily as it does from the area with lower solute concentration, so there is net fluid movement in one direction;

- Water passes from an area of high potential (i.e. low solute concentration) to an area of lower potential (higher solute concentration). That is to say, water moves to the compartment with the lower solvent concentration.

These models are not mutually exclusive but merely express the process in different ways; solute concentration gradients 'drive' osmotic effects. Whichever model we prefer to use, the outcome is the same: equality of solute *concentration* is achieved at the expense of change, and therefore inequality, of *volumes*.

Strictly speaking, osmotic pressure is the pressure that needs to be applied to prevent osmosis occurring, not, as is often assumed, the pressure 'pulling' water across the membrane. An alternative term which is often used is 'osmotic potential', implying the possibility of an osmotically-driven redistribution of solvent. The magnitude of osmotic potential is defined by: (1) the total concentration (expressed in mmol) of all osmotically active solutes of the solution, per *kilogram* of solvent, and termed the osmolality; and (2) the nature (i.e. permeability) of the separating membrane. The *effective* osmotic potential that excludes the presence of any freely diffusible molecules or ions is termed the 'tonicity'. Two solutions of the same osmolality are by definition iso-osmotic, but they may not be isotonic if their chemical compositions are qualitatively markedly different. For example, let us suppose one solution is composed of small molecules which may cross the membrane easily, whilst the other solution contains an equivalent concentration of a non-diffusible solute. The non-diffusible solute would exert a greater osmotic effect, causing an unequal distribution of liquid in the two compartments. In this situation, we might expect the diffusible solute to be unequally distributed across the membrane in order to 'compensate' for the presence in one compartment of the non-diffusible solute, thus the result would be two solutions of different volumes *and* different chemical composition (see Figure 1.10).

Cell membranes, with the notable exception of those in parts of the renal nephron, are essentially permeable to water due to the presence of pores called aquaporins. The extent to which fluid distribution occurs physiologically is determined *not* by the total solute

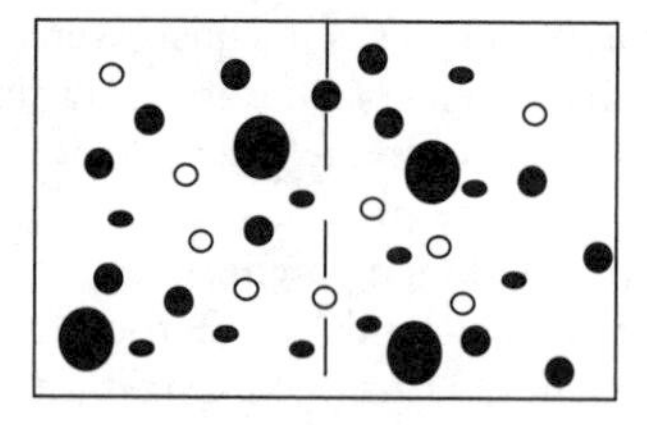

Iso-osmotic and isotonic. Both compartments have the same total number of (small) diffusible and (large) non-diffusible particles.

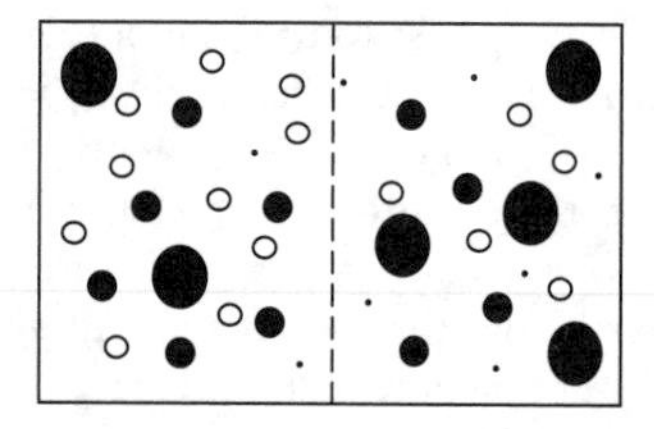

Iso-osmotic (same total number of particles) but *not* isotonic as the two compartments have different numbers of diffusible particles.

Figure 1.10 Tonicity: effective osmotic pressure arising from the net effects of diffusible and non-diffusible solutes

concentration (measured in the clinical laboratory as osmolality), but by the net concentration of the membrane-impermeable solutes present in the intracellular and extracellular fluids, i.e. the tonicity of the ICF and ECF.

At 4°C, 1 litre of water weighs 1 kg so 1 mmol/kg = 1 mmol/L, i.e. osmolality = concentration, but at body temperature the density of water is approximately 0.992 g/mL, so 1 kg of water is more than 1 litre, so osmolality greater than concentration. If in a kilogram of solvent we have 1 mole of solute the solution is, by definition, 1 molal; if, however, the solute dissociates into two particles (e.g. NaCl), 1 mole solute/kg equates to a 2 molal solution.

(b) The Gibbs-Donnan equilibrium

The situation described above applies adequately to small, uncharged solutes, but the presence of electrolytes in body fluids introduces another factor influencing solute and solvent distribution across a cell membrane. The Gibbs-Donnan law of ionic equilibria requires that when two solutions are separated by a permeable membrane, the net charge in the two compartments must be the same, i.e. there must *not* be an electrical gradient (refer to Figure 1.2 in Section 1.i). If a membrane separates solutions contained in two compartments, identified as A and B for convenience, and electrolyte solutions of different concentrations are placed into A and B, then at equilibrium two criteria will be met:

(1) Each solution must be electrically neutral, so

$$[\text{cations}]_A = [\text{anions}]_A$$

and

$$[\text{cations}]_B = [\text{anions}]_B$$

(2) The product of the concentrations of *diffusible* ions on each side are equal. In simple mathematical terms this is:

$$\begin{array}{ccc} [\text{cations}]_A \times [\text{anions}]_A & = & [\text{cations}]_B \times [\text{anions}]_B \\ \text{total ionic charge in} & = & \text{total ionic charge in} \\ \text{compartment A} & & \text{compartment B} \end{array}$$

Thus, for simple solutions of sodium chloride we can write;

(1)
$$[Na^+]_A = [Cl^-]_A$$

and

$$[Na^+]_B = [Cl^-]_B$$

(2)
$$[Na^+]_A \times [Cl^-]_A = [Na^+]_B \times [Cl^-]_B$$

The ICF and ECF are, however, really colloidal solutions, meaning that they contain proteins which are large poly-electrolytes, i.e. molecules that carry multiple electrical charges, and are, moreover, non-diffusible across a cell membrane. The ICF also contains a high concentration of non-diffusible organic phosphate anions such as ATP. The presence of these charged species with the diffusible ions such as Na^+ and Cl^- disturbs the simple equilibrium described above, creating an unequal distribution of mobile ions across the membrane.

At the normal, and slightly alkaline, physiological pH of blood, proteins collectively carry a net negative charge of approximately 12 mEq/L; this charge will tend to sequester positively charged ions

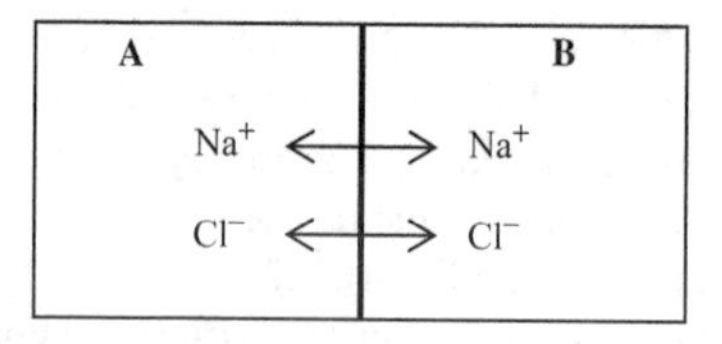

Only diffusible ions are present;
Concentrations of Na^+ and Cl^- in each compartment are equal
total positive and negative charges are equal in A and B.

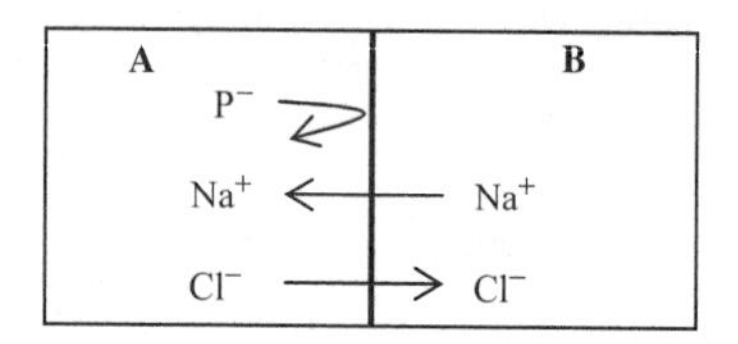

Compartment A contains a non-diffusible anion (P^-), e.g. protein.
Concentration gradient exists; $[Na^+]_A > [Na^+]_B$; $[Cl^-]_A < [Cl^-]_B$
but electroneutrality is maintained in presence of P^-.

Figure 1.11 Gibbs-Donnan equilibrium

such as sodium. If protein and organic phosphate (together represented as P^- in Figure 1.11) are present in only one compartment (say, A), there will be an imbalance in the distribution of ions between the two compartments. The consequence is that, at equilibrium, the actual concentration of Na in the protein-containing compartment (A) will be slightly higher than in the compartment without protein (B), whilst the Cl^- concentration in B will be lower; in effect, the negatively charged protein has 'displaced' some of the chloride to ensure electrical neutrality.

In reality, protein is present on both sides of physiological membranes; however, ISF is protein-poor relative to both ICF and plasma. Furthermore, qualitative differences in the protein content and any pH differences across the boundaries will lead to protein-based charge differences, thus we would expect Gibbs-Donnan effects to occur at two interfaces: plasma/ISF and ISF/ICF. Overall, therefore, any abnormality that leads to changes in protein distribution between compartments or an acid-base disturbance affecting charge density across a membrane may have an impact on the flux of fluid and electrolytes between compartments.

Referring back to the two criteria we met earlier:

(1) Electrical neutrality within each compartment is achieved when

$$[Na^+]_A = [Cl^-]_A + [P^-]_A \qquad \ldots \text{(i)}$$

and

$$[Na^+]_B = [Cl^-]_B \qquad \ldots \text{(ii)}$$

note therefore that $[Na^+]_A > [Cl^-]_A$ ... (iii)

(2) $[Na^+]_A \times [Cl^-]_A \times [P^-]_A = [Na^+]_B \times [Cl^-]_B$ ... (iv)
Given that $[Na^+]_A > [Cl^-]_A$ (see (iii) above), for (iv) to be satisfied, the following must be true;

$$[Na^+]_A > [Na^+]_B$$

and

$$[Cl^-]_A < [Cl^-]_B$$

Overall, therefore, $[Na^+]_A + [Cl^-]_A > [Na^+]_B + [Cl^-]_B$

Expressed simply, this reveals that *concentration gradients* now exist for both Na^+ and Cl^- across the membrane. The compartment containing the non-diffusible ion (P^-, proteins and organic phosphates) has a higher total solute concentration (osmolality) than compartment B, and therefore an osmotic gradient exists across the membrane.

SECTION 1.xii

Water and ion distribution between compartments 1: Physiology

The account of fluid and ion distribution given above in Section 1.xi describes an *in vitro* situation where a synthetic membrane is the boundary between two solutions, but the Gibbs-Donnan phenomenon has physiological significance.

Cell membranes are selectively permeable barriers; water and small hydrophobic molecules are able to cross the membranes but charged molecules cannot cross unaided. If we were to make an experimental artificial cell membrane and use this to separate two physiological fluids and study the fluxes between compartments, we would find that, despite their charge, ions such as Na^+ and K^+ are fairly freely diffusible across cell membranes and eventually an equal distribution of the ions would be established. The ion concentration gradients between ICF and ECF evident in Table 1.3 are 'unstable' and so must be maintained by active transport mechanisms associated with the plasma membrane of cells. Such active transport processes are frequently referred to as ion 'pumps'.

Ion pumps

When we consider the chemical composition of ICF and interstitial fluid (ISF) we note that the major difference is that ISF is essentially

Essential Fluid, Electrolyte and pH Homeostasis, First Edition. Gillian Cockerill and Stephen Reed.

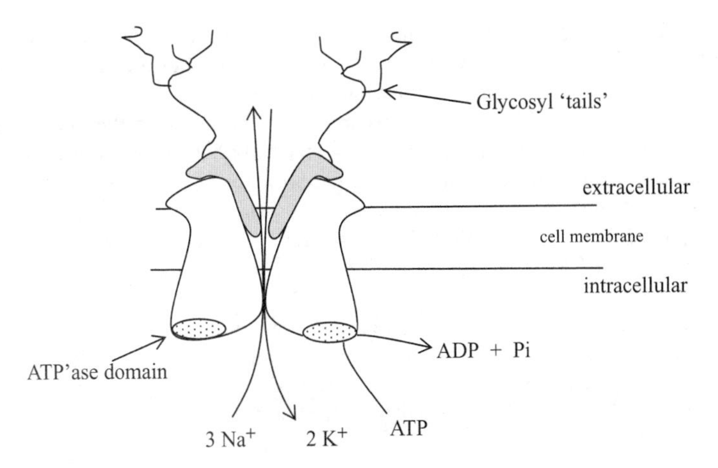

Figure 1.12 Sodium–potassium dependent ATP'ase ('sodium pump')

phosphate and protein-free, whilst the ICF is protein and phosphate-rich. If a cell membrane acted simply as a barrier between the ICF and ISF, we would expect a net flow of water *into* a cell due to the higher osmolality of that compartment. The cell would swell and eventually burst. The fact that this does not happen is because the typical cell membrane contains ion transport mechanisms ('pumps') which actively extrude or exchange ions. The sodium pump, strictly called the Na^+/K^+ ATP'ase (see Figure 1.12), is just such a mechanism. A significant proportion of the adenosine triphosphate (ATP) generated by a cell is used to drive the sodium pump to maintain appropriate ion compositions in the ICF and ISF (see Table 1.3 in Section 1.viii).

The sodium-potassium dependent ATP'ase is found in all tissues, and functions to move sodium out of cells in exchange for potassium which enters the cell. For every 3 sodium ions which are extruded, 2 potassium ions are taken in. This is an active process, i.e. one that requires a constant supply of energy, supplied by ATP. If the availability of metabolic fuels such as glucose or fatty acids becomes restricted, the pump mechanism fails and the ion concentrations will eventually equilibrate across the cell membrane. Such a situation rarely occurs *in vivo*, but does occur in blood samples awaiting analysis in the laboratory. If blood samples are not received promptly in the laboratory and the plasma separated from cells, the Na/K pump very soon begins to fail and potassium leaks out of red cells down the concentration gradient that exists between the ICF and ISF, and thus measurement of potassium concentration for clinical purposes is invalidated.

Similar ATP-driven ion pumps are also to be found in the kidney (particularly the distal tubule) and stomach (parietal cells). Cells located throughout the length of the nephron have ion pumps located on both their luminal (also called apical) surface in contact with the glomerular filtrate, and on their basolateral surface allowing ions to enter the ECF. These pumps, which are essential for the reabsorption of sodium and secretion of potassium and hydrogen ions, are often regulated by hormonal action. Many of these pump mechanisms are, like the Na^+/K^+ ATP'ase, antiports meaning that two types of ion are transported simultaneously in opposite directions across the membrane. There are many particular examples of such pumps, invariably found at specific sites along the nephron but notably in the proximal tubule and the collecting duct and which are responsible for normal ion homeostasis and acid-base balance. Examples of such transporters include the sodium/hydrogen exchanger (NHE-3), the ammonium/potassium antiporter, the Na-K-Cl_2 (NKCC) co-transporter and the electroneutral anion exchanger (AE1). Interestingly, the AE1 is structurally homologous with a protein called 'band-3' found in the membrane of red blood cells, where it too plays a role in acid-base regulation via its key role in HCO_3^- translocation.

Gastric parietal cells are stimulated by the hormone gastrin to secrete hydrochloric acid into the lumen of the stomach. Parietal cells, like red cells and kidney cells, are rich in the enzyme carbonic anhydrase which promotes the formation of carbonic acid from carbon dioxide and water. Dissociation of carbonic acid liberates H^+ and HCO_3^- ions; the protons pass into the cavity of the stomach as gastric acid while the bicarbonate ions enter the bloodstream.

Calcium concentration in ECF is approximately 2.5 mmol/L in health, but the intracellular concentration of ionised Ca^{2+} is much lower (Table 1.3). Increases in ICF Ca^{2+} concentration initiate a number of tissue-dependent effects, such as neurotransmitter release from neurones or muscle contraction. One mechanism whereby ICF $[Ca^{2+}]$ will rise is by the opening of calcium channels in the plasma membrane; effectively a valve opens and calcium ions flood into the cell. Subsequently, calcium must be pumped out of the cell (or compartmentalised within the cell) to allow the 'resting' $[Ca^{2+}]$ to be regained.

It is clear therefore that correct physiological regulation of ion movements and ion concentration is critical to normal functioning of the human body. Disturbances to the pump mechanisms can be dangerous. For example, cystic fibrosis is the commonest inherited metabolic defect (IMD) in European ethnic groups. One of the symptoms of this condition is accumulation of mucus in the airways and ducts within the body, and this arises because of a defect in a chloride transporter (called CFTR) located in the cell membrane. The inability to pump chloride ions upsets the osmotic balance and results in the production of an abnormally sticky secretion, causing blockage of secretory ducts and promoting infection.

Water distribution

As shown in Figure 1.13, there are two barriers to fluid flow between compartments: (a) the wall of the blood vessel separating the plasma

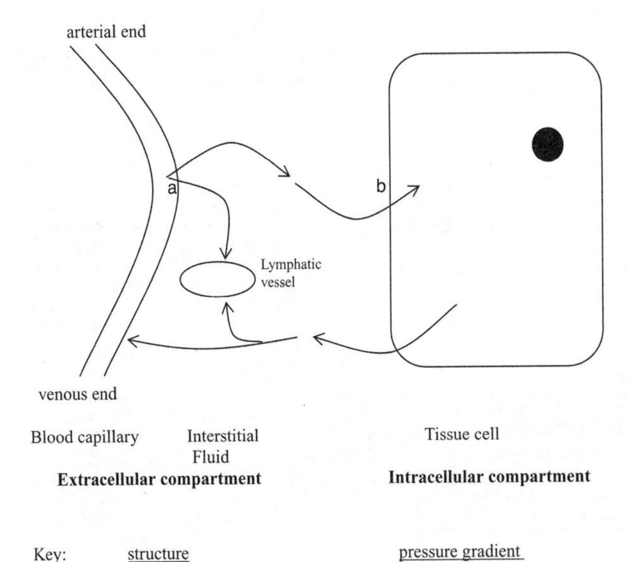

Figure 1.13 Fluid and solutes cross two barriers between plasma and the interior of a cell

from the ISF; and (b) the external membrane of each cell separating its ICF from the ISF.

Movement of water from plasma to the ISF is determined by the net effects of the blood hydrostatic pressure, itself a function of blood pressure, pushing fluid out of the vessel, opposed by protein gradient pulling water back into the vessel. The effect of the protein, principally albumin, in this manner is referred to as oncotic pressure (Figure 1.13).

Our current understanding of fluid exchange between the plasma and the ISF are described in pressure terms by Starling's hypothesis (Figure 1.14). The blood vessel endothelial wall is freely permeable to small molecules, but essentially impermeable to proteins, thus the ISF has a very low protein concentration relative to the plasma. Movement of fluid across the vessel wall is determined by the balance of two opposing forces: hydrostatic pressure and oncotic pressure due to the protein content of the fluids. Blood pressure and therefore hydrostatic pressure is lower at the venous end of a capillary than at the arterial end because of the wider internal diameter of a venule compared with an arteriole, so here the inward-acting plasma protein oncotic pressure exceeds the hydrostatic pressure, and fluid now moves from the ISF into the blood vessel. The importance of Starling's hypothesis is explained more fully in Section 2.i.

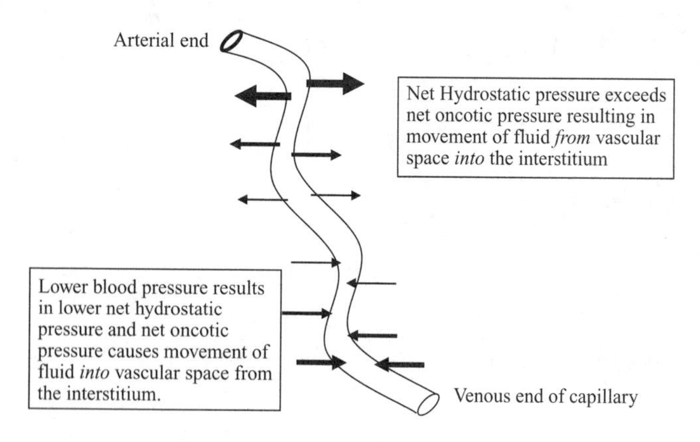

Figure 1.14 Starling's hypothesis: fluid movement into and out of blood capillary

SECTION 1.xiii

Osmoregulation: solvent and solute balance

The brief account of Starling's hypothesis given in the previous section describes the movement of fluid from the blood plasma into the interstitial fluid. This Section gives an account of the passage of fluid between the ISF and the intracellular compartment.

The internal fluid fluxes across cell membranes are driven by osmotic forces. Osmosis is the flow of solvent (water) across a selectively permeable membrane from an area of lower solute content to an area of higher solute content in order to equalise the pressures. When two solutions have equal *effective* osmotic potential (isotonic), there is no *net* flow of liquid between them; rather, a steady-state equilibrium is established as fluid flows in both directions at the same rate. Note that terms such as isotonic and iso-osmotic are relative, measured against the physiological force acting upon cell membranes tending to cause the cell to swell (if inside pressure is greater than that outside) or shrink (when inside force is lower than that outside). The normal pressure is exerted when osmolality is approximately 290 mmol/kg.

The true osmotic potential of a solution is determined by the total number of all solute particles present. The principal osmotically-active solutes found in body fluids are sodium (controlling water flow between ICF and ISF) and protein (controlling fluid flow between the plasma and the ISF). The effect of protein (albumin being quantitatively the most significant) is often called oncotic pressure. Providing that the osmolalities of the ECF and ICF are balanced, cells will be in an appropriate state of hydration (Figure 1.15).

Essential Fluid, Electrolyte and pH Homeostasis, First Edition. Gillian Cockerill and Stephen Reed.

(a) normal steady state

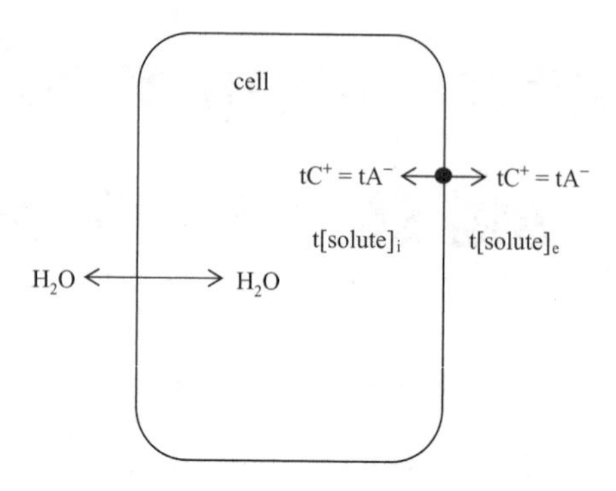

Key:

$t[solute]_i$	intracellular total solute concentration
$t[solute]_e$	extracellular total solute concentration
tC^+	total number of positive charges (cations)
tA^-	total number of negative charges (anions), includes proteinate, phosphate and organics *etc.*

● ion transport protein   diffusion ⟶

(b) hypotonic ISF = hypertonic ICF

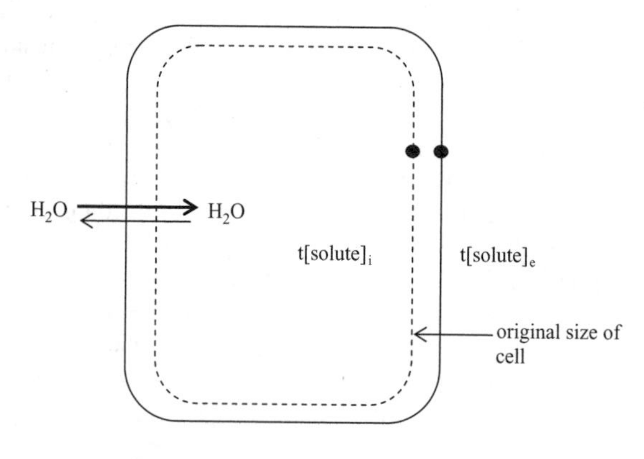

$t[solute]_i > t[solute]_e$
Net flow of water into cell = swelling (overhydration)

Figure 1.15 Osmotic effects on cell volume

(c) hypertonic ISF = hypotonic ICF

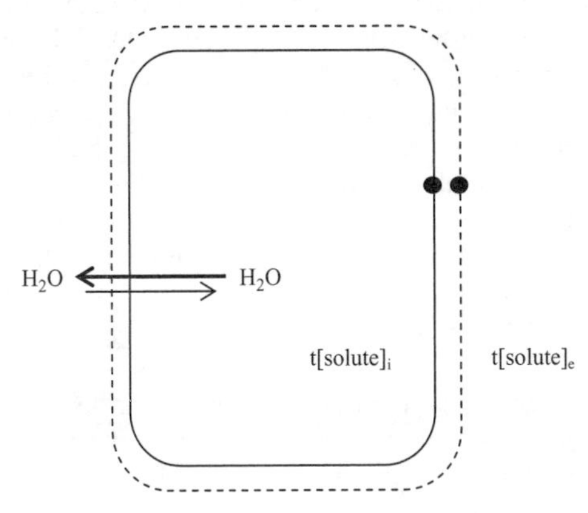

t[solute]$_i$ < t[solute]$_e$
Net flow of water out of cell = shrinkage (dehydration)

Figure 1.15 (*Continued*)

Three similar terms, which are often confused, are all used to indicate the osmotic potential of a solution:

(1) *Osmolarity*: expressed in units of mmol per *litre* of water.

This calculated value is an estimate based on measurement of the concentration of the most significant ions and molecules in plasma or urine. Many equations have been developed to calculate osmolarity from measured parameters. For example:

(i) $\text{Osm} = 2 \times [\text{Na}]$
(ii) $\text{Osm} = 2 \times [\text{Na}] + 10$
(iii) $\text{Osm} = 2 \times ([\text{Na}]+[\text{K}])$
(iv) $\text{Osm} = 2 \times ([\text{Na}]+[\text{K}]) + [\text{urea}]$
(v) $\text{Osm} = 2 \times ([\text{Na}]+[\text{K}]) + [\text{urea}] + [\text{glucose}]$
(vi) $\text{Osm} = 1.78 \times ([\text{Na}]+[\text{K}]) + [\text{urea}] + [\text{glucose}]$
(vii) $\text{Osm} = 2[\text{Na}] + 0.36[\text{urea}] + 0.06[\text{glucose}]$

(2) *Osmolality*: expressed in units of mmol per *kg* of water.

This is a measured parameter that assesses the *activity* of all solutes present in a sample of plasma or urine. As such, osmolality is a better measure of the true physiological condition than is osmolarity because (i) it takes into account a wider range of solutes and (ii) the movement of fluid between physiological compartments is determined by the thermodynamic *activity* of the solutes rather than just their concentration. Osmolality is a function of the number of solute molecules present in the sample, not their size or nature. Proteins have only little effect on osmolality because, despite their high concentration, there are relatively few individual molecules present in comparison with, say, sodium ions. Note also that all solutes in a sample of plasma will contribute to the measured osmolality, but not all of the solutes will exert a physiological effect in the body, so osmolality is not the true physiologically important parameter. This point is clarified in the following account of tonicity.

Depending upon which formula is used to derive a value for osmolarity, the measured osmolality is usually approximately 10 mmol/L higher than the calculated osmolarity for the same sample, i.e.:

$$\{\text{measured osmolality}\} - \{\text{calculated osmolarity}\} = \sim 10\text{–}12\ \text{mmol/L}$$

This numerical difference, called the 'osmolal gap', arises due to the fact that only selected constituents are used when calculating osmolarity. Increases in the value of the osmolal gap indicate the presence in plasma of other, implicitly unmeasured, osmotically important solutes. It is important to note that the osmolal gap is not a true physiological phenomenon but merely a reflection of differences in laboratory measurements employed to assess osmotic potential of samples.

In health, the osmolality of plasma (ECF) is 285–295 mmol/kg. The osmolality of urine varies widely according to recent fluid intake, but is typically 2–3 times higher than the osmolality of plasma.

Osmolarity and osmolality are laboratory-measured parameters that reflect the total solute concentration or total solute activity in solution. Both parameters are used to give an estimate of tonicity.

(3) *Tonicity*: the true *physiologically effective* osmotic pressure, i.e. the physiological force acting upon cell membranes.

Only non-diffusible solutes can have such an effect and the principal contributor to tonicity is sodium, therefore it is this ion which primarily determines the state of cellular hydration. *In vivo*, not all solutes exert an osmotic effect. Urea, for example, is not osmotically active in ECF because it diffuses freely across cell membranes; glucose, in the presence of adequate insulin activity, will enter cells easily so it too is not *normally* osmotically significant. Tonicity is not measured directly by routine laboratory tests, but it is in effect the numerical difference between osmolality and the plasma concentrations of glucose and urea. Rapid changes in blood glucose or urea concentrations may exert a temporary osmotic influence in the ECF, but both solutes will eventually become distributed equally between the ICF and ECF compartments. Only if plasma glucose concentration is elevated for a long period of time will it affect fluid distribution between ISF and ICF.

It has been estimated that the tonicity of ICF is actually slightly higher than that of ISF. In order to prevent cells swelling and bursting, ions, and thereby water, are actively extruded via pump mechanisms. A significant proportion (estimated to be between 20% and 40%) of the daily ATP generation of cells is consumed in trying to regulate the ionic gradient to prevent damage to the cells; this is so-called 'osmotic work'.

SECTION 1.xiv

Self-assessment exercise 1.3

1. What is the volume (in litres) of each of the following?
   (i) total body water
   (ii) ECF
   (iii) ICF
   (iv) interstitial fluid
   in (a) a typical 70 kg man
   and (b) an 85 kg obese female (assume 1 litre of water weighs 1 kg).
   NB: you will need to make some reasonable assumptions to answer this question.
2. What is the total amount (in mmol) of (i) Na^+ and (ii) K^+ in body fluids of a 70 kg adult male?
3. Estimate the total number (amount) of osmotically active solutes in (i) ECF and (ii) ICF.
4. Two solutions, A and B, are separated by a selectively permeable membrane. A and B are iso-osmotic and isotonic with respect to each other and both contain a mixture of large (non-diffusible) and small (diffusible) solutes and equal concentrations of anions and cations. Assume that A and B have the same volume, say 2 litres.
   State with reasons what effect the following would have on the relative volumes and chemical compositions of A and B if:
   (i) 500 mL of isotonic liquid containing only small *diffusible non-electrolytes* was added to A.
   (ii) 500 mL of isotonic liquid containing only small *diffusible electrolytes* (both anions and cations) was added to A.
   (iii) 500 mL of isotonic liquid containing only large *non-diffusible* non-electrolytes was added to A.

Essential Fluid, Electrolyte and pH Homeostasis, First Edition. Gillian Cockerill and Stephen Reed.

(iv) 500 mL of isotonic liquid containing only large *non-diffusible* anionic electrolytes was added to A.
(v) 500 mL of isotonic liquid containing a mixture of small diffusible and large non-diffusible solutes was added to A.
(vi) 500 mL of hypertonic liquid containing only small diffusible solutes including electrolytes was added to A.

5. So-called 'normal' physiological saline is prepared by dissolving NaCl in water at a final concentration of 0.85%.
   (i) What is the osmolality of this solution?
   (ii) What effect would there be on red cells suspended in 0.45% saline?
6. Verify that the average charge carried by plasma proteins at normal pH is approximately 12^-. For simplicity consider only the major ions and assume that the charge on phosphate is 2 mEq/L and organic ions ~8 mEq/L. *(HINT: Refer to Table 1.3 and Fig. 1.8.)*
7. Calculate (i) the osmolal gap and (ii) the anion gap for the following results:

Measured parameter	Subject 1	Subject 2
Na^+ mmol/L	135	141
K^+ mmol/L	3.8	4.2
Cl^- mmol/L	102	98
HCO_3^- mmol/L	20	27
Urea mmol/L	4.2	4.1
Glucose mmol/L	4.8	–
Osmolality mmol/kg	298	315

8. At what temperatures would (i) a sample of urine with an osmolality of 450 mmol/kg and (ii) a plasma sample with an osmolality of 275 mmol/kg freeze?

Check your answers *before* continuing to Part 2 of the text.

SECTION 1.xv

Summary of Part 1

- Body fluids are complex mixtures of solutes.
- Physical and chemical properties of water make it an ideal solvent, and the colligative properties of water are a function of the solute content.
- pKa and the Henderson-Hasselbalch equation provide the basis for an understanding of acids, bases and buffers.
- 60–70% of body weight is water; ~42 litres in a 70 kg adult male.
- Two-thirds of the total volume of body fluids is intracellular fluid; one third is extracellular (i.e. plasma and interstitial fluid, ISF).
- There is exchange of water and solutes between ECF, ISF and ICF.
- Specific chemical compositions of the ICF, ISF and plasma are significantly different, but the osmotic potentials and overall charges are comparable.
- Chemical and ionic (charge-related) equilibria operate across cell membranes.
- Maintenance of osmotic balance and electrical neutrality are key features of homeostatic control.
- Starling's hypothesis predicts the movement of fluid between the plasma and the ISF according to relative hydrostatic forces (blood pressure) and oncotic (protein-related) pressure gradients.
- Water movement across cell membranes is a function of osmotic gradients, whilst Gibbs-Donnan equilibria influence the distribution of ions.
- Tonicity of body fluids determines the movement of water between the ISF and the ICF across cell membranes.
- Na concentration is the major determinant of tonicity and therefore water movement between ISF and ICF.
- Osmolality and osmolarity are reasonable estimates of the true tonicity of plasma or urine.
- The osmolal gap and the anion gap do not exist *in vivo* but are useful derived parameters.

Essential Fluid, Electrolyte and pH Homeostasis, First Edition. Gillian Cockerill and Stephen Reed.

Answers to Part 1 self-assessment exercises

Self-assessment exercise 1.1

1. Calculate the pH of pure water at 18°C and 40°C.

Neutral pH of pure water at 18°C

$$pH = -\log_{10}[H^+]$$

$$\text{at } 18°C,\ K_w = [H^+] \times [OH^-] = 0.64 \times 10^{-14}$$

$$\therefore [H^+] = \sqrt{0.64 \times 10^{-14}}$$

$$= 8 \times 10^{-8}$$

$$pH = -\log_{10}(8 \times 10^{-8})$$

$$\log 8 \times 10^{-8} = -7.097$$

$\therefore$ the *negative* log of $8 \times 10^{-8} = 7.097$

thus, $pH = 7.10$

similarly,

Neutral pH of pure water at 40°C

$$pH = -\log_{10}[H^+]$$

$$\text{at } 40°C,\ K_w = [H^+] \times [OH^-] = 2.92 \times 10^{-14}$$

$$\therefore [H^+] = \sqrt{2.92 \times 10^{-14}}$$

Essential Fluid, Electrolyte and pH Homeostasis, First Edition. Gillian Cockerill and Stephen Reed.

$$= 1.709 \times 10^{-7}$$
$$\text{pH} = -\log_{10}(1.709 \times 10^{-7})$$
$$\therefore \text{pH} = 6.77$$

2. From your answer to (a) above, what can you surmise about pH neutrality at body temperature?

   Given that body temperature is 37°C, i.e. close to 40°C, we would expect that neutrality in our cells and tissues is *not* pH 7 but approximately 6.77. In fact the estimated pH of ICF is 6.85. The lower pH is *not* an indication that cellular cytosol is acidic, but a reflection of the different extent of dissociation of water molecules at body temperature compared with 25°C.

3. Normal arterial whole blood pH falls in the range 7.35 to 7.45. Express these two values in nmol/L hydrogen ion concentration (1 nmol = 10^{-9} mol).

$$\text{pH} = -\log[\text{H}^+]$$
$$\log[\text{H}^+] = (-\text{pH})$$
$$[\text{H}^+] = \text{antilog}(-\text{pH})$$
$$\text{if pH} = 7.35,\ \text{antilog}(-7.35) = 0.00000004467\ \text{mol/L}$$
$$\times 10^9 \text{ to convert mol/L to nmol/L} = 44.7\ \text{nmol/L}$$

similarly,

$$\text{pH} = -\log[\text{H}^+]$$
$$\log[\text{H}^+] = (-\text{pH})$$
$$[\text{H}^+] = \text{antilog}(-\text{pH})$$
$$\text{if}\quad \text{pH} = 7.45,\ \text{antilog}(-7.45) = 0.00000003548\ \text{mol/L}$$
$$\times 10^9 \text{ to convert mol/L to nmol/L} = 35.5\ \text{nmol/L}$$

4. The measured osmolality of a plasma sample was found to be 295 mOsm/L. Calculate the actual depression of freezing point as measured by the analyser.

   If a 1 osmolal (= 1000 milliosmolal) solution depresses the freezing point by 1.86°C, then by simple proportions:

$$\frac{295}{1000} \times 1.86 = 0.549°\text{C}$$

5. Study the table below.

Molecular property	CH_4	NH_3	H_2O
Molecular weight	16	17	18
Freezing point °C	−180	−80	0
Boiling point °C	−160	−35	100
Relative viscosity	0.1	0.25	1

How can the differences revealed above be explained?

The values in the table reflect the relative 'cohesiveness' of the molecules, i.e. their ability to interact with each other. Despite the similarities in size and structure, the unique polarity of water confers particular properties. Water molecules, being very polar, are able to attract each other strongly, whereas CH_4 molecules interact very weakly with each other.

6. Given that a normal plasma [K] is approximately 4.0 mmol/L and a typical plasma glucose concentration is 5.0 mmol/L, how many mg of K and glucose are there in 100 mL of plasma?

$$4.0\ \text{mmol/L} = 0.4\ \text{mmol/100 mL}$$

by definition, $$1\ \text{mmol K} = 39\ \text{mg}$$

$$0.4\ \text{mmol K} = 15.6\ \text{mg/100 mL}$$

$$5.0\ \text{mmol/L} = 0.5\ \text{mmol/100 mL}$$

by definition, $$1\ \text{mmol glucose} = 180\ \text{mg}$$

$$0.5\ \text{mmol glucose} = 90\ \text{mg/100 mL}$$

Note that this calculation emphasises the value of the SI system when expressing concentrations. Two solutes of approximately equal molarity (which is the physiologically important parameter) may have very different mass concentrations.

7. Express 140 mmol/L sodium as mEq/L and as mg/L.

Sodium is monovalent so

$$1\ \text{mmol} = 1\ \text{mEq} \quad 140\ \text{mmol/L} = 140\ \text{mEq/L}$$

$$1\ \text{mmol Na} = 23\ \text{mg}$$

$$140\ \text{mmol Na} = 3220\ \text{mg}$$

$$140\ \text{mmol/L} = 3220\ \text{mg/L}$$

8. Express 1.5 mg/100 mL of creatinine ($C_4H_7N_3O$) in SI units.

Molecular weight of creatinine = 113
1.5 mg/100 mL = 15 mg/L

$$\frac{15}{113} = 0.133 \text{ mmol/L} = 133 \text{ µmol/L}$$

9. If a sample of plasma has a calcium concentration of 2.5 mmol/L and a potassium concentration of 136.5 mg/L, what are these concentrations expressed in mEq/L?

1 mmol/L calcium = 2 mEq/L
2.5 mmol/L Ca^{2+} = 5.0 mEq/L

$$136.5 \text{ mg/L K} = \frac{136.5}{39} = 3.5 \text{ mmol/L} = 3.5 \text{ mEq/L}$$

10. A sample of urine freezes at −0.84°C. What is the molality of this sample?

If a 1 molal (1000 millimolal) solution freezes at −1.86°C, then by simple proportions a solution which freezes at −0.84°C will be 0.451 molal (= 451 mmolal);

$$1000 \times \frac{-0.84}{-1.86} = 451 \text{ mmolal}$$

11. Convert (i) 40 mmHg to kPa, and (ii) 13.0 kPa to mmHg.

$40 \times 0.133 = 5.3$ kPa
$13/0.133 = 97.7$ mmHg

These values are physiologically significant; 5.3 kPa and 13 kPa are typical values for the pressure of CO_2 and O_2 respectively in arterial blood.

Are your answers correct? If so, well done and return to Section 1.v; If your answers are not correct, re-read Sections 1.i to 1.iv before continuing.

Self-assessment exercise 1.2

1. Complete the following table by calculating values indicated by ??

   Arrange the acids as a list with the strongest at the top and weakest at the bottom.

Acid	K_a	pK_a
Acetic	1.83×10^{-5}	4.74
Carbonic	7.9×10^{-7}	6.1
Acetoacetic	2.6×10^{-4}	3.58
Fumaric	$Ka_1 = 9.3 \times 10^{-4}$	3.03
	$Ka_2 = 2.88 \times 10^{-5}$	4.54
Phosphoric	$Ka_2 = 1.6 \times 10^{-7}$	6.796 (~6.8)
Succinic	$Ka_1 = 6.6 \times 10^{-5}$	4.18
	$Ka_2 = 2.75 \times 10^{-6}$	5.56
Pyruvic	3.16×10^{-3}	2.50
Aspartic side chain COOH	1.26×10^{-4}	$pKa_2 = 3.9$
Histidine	1×10^{-6}	$pKa_3 = 6.0$

(a) Derivation of pK_a from K_a

For example, acetic acid: $pK_a = -\log K_a$

given that $pK_a = -\log 1.83 \times 10^{-5}$

$\log 1.83 \times 10^{-5} = -4.74$

$-\log 1.83 \times 10^{-5} = 4.74$

and carbonic acid: $pK_a = -\log K_a$

given that $K_a = -\log 7.90 \times 10^{-7}$

$\log 7.90 \times 10^{-7} = -6.10$

$-\log 7.90 \times 10^{-7} = 6.10$

(b) Derivation of K_a from pK_a

$K_a = \text{antilog}(-pK_a)$

For aspartic acid $K_a = \text{antilog}(-3.9)$

$\text{antilog} - 3.9 = 0.0001259$

$= 1.26 \times 10^{-4}$

An alternative way to make this calculation is to take the antilog of difference between the pK_a and the next nearest whole number.

For example, pyruvic: $pK_a = 2.5$ so the next nearest whole number is 3

Then $3 - 2.5 = 0.5$ antilog $0.5 = 3.16$
Because the next nearest whole number is 3 we raise the antilog to 10^{-3}
So the answer $= 3.16 \times 10^{-3}$

The strongest acid in this list is pyruvic acid (lowest pK_a)

followed by	fumaric acid (pKa_1)
	acetoacetic acid
	aspartic acid,
	succinic acid (pKa_1)
	fumaric acid (pKa_2)
	acetic acid
	succinic acid (pKa_2)
	histidine
	carbonic
and the weakest acid is	phosphoric (pKa_2, highest pK_a)

2. How does knowledge of the ionic behaviour of amino acids help us understand the buffer action of proteins?

   A protein is a polymer of individual amino acids arranged in such a way that the side chains ('R' groups) of the amino acids project away from the backbone structure:

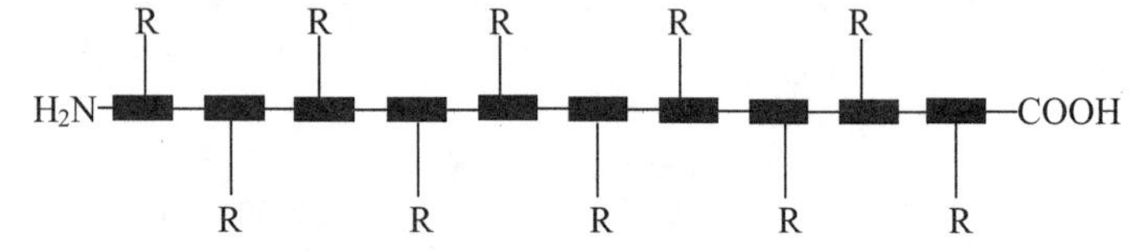

Those R groups that include an amino group (e.g. lysine) or a carboxylic acid group (e.g. aspartate) are subject to protonation or deprotonation, and thus act as weak conjugate acids (proton donors: $COOH = COO^- + H^+$ or $NH_3^+ = NH_2 + H^+$) or weak bases (proton acceptors: $COO^- + H^+ = COOH$ or $NH_2 + H^+ = NH_3^+$).

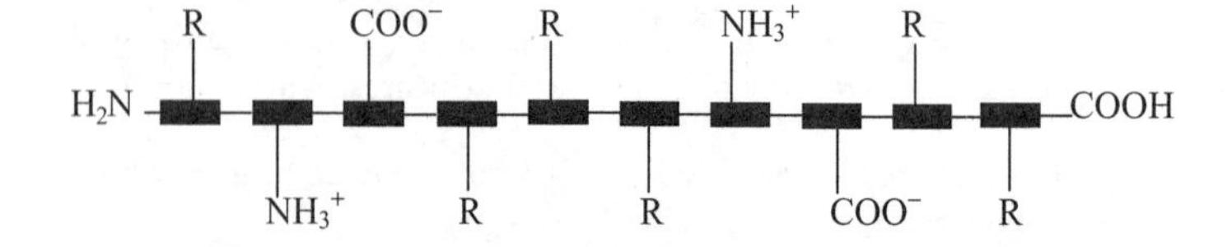

Thus, by accepting or donating protons the $[H^+]$ of the surrounding fluid can be regulated (buffered).

3. Write balanced chemical equations to show the dissociation of pyruvic acid ($CH_3.CO.COOH$) and of phosphoric acid (H_3PO_4).

Pyruvic acid $CH_3.CHOH.COOH \rightleftarrows CH_3.CHOH.COO^- + H^+$

Phosphoric acid $H_3PO_4 \rightleftarrows H_2PO_4{}^- + H^+$ pKa_1

$H_2PO_4{}^- + H^+ \rightleftarrows HPO_4{}^{2-} + H^+$ pKa_2

$HPO_4{}^- + H^+ \rightleftarrows PO_4{}^{3-} + H^+$ pKa_3

4. State, giving reasons, whether (i) histidine (imidazole side chain $pK_a = 6.0$), and (ii) the side chain amino group of arginine ($pK_a = 12.6$) would be protonated or deprotonated at typical cytosolic pH of 6.85.

If pH of the surrounding fluid is greater than the pK_a, the group will be protonated (i.e. act as a base). This is the case for histidine ($6.85 > 6.0$). Because the difference between the pK_a and the pH is less 1 pH unit, which is within the buffering range of histidine imidazole side chain group, this amino acid is an important physiological protein-associated buffer.

When pH of the surrounding fluid is less than pK_a, the group will be deprotonated, (i.e. act as an acid). The side chain NH_2 of arginine has a pK_a greater than 6.85, so it will be deprotonated.

You may wish to remind yourself of ionic properties of amino acids by reading the relevant chapter in a biochemistry textbook.

5. What proportions (ratio) of lactic acid and sodium lactate would be required to prepare a buffer with a final pH of 3.5? (pK_a lactic acid = 3.8)

To solve this problem we need to use the Henderson-Hasselbalch equation.

$$pH = pK_a + \log\{[base]/[acid]\}$$

$$3.5 = 3.8 + \log\{[lactate]/[lactic\ acid]\}$$

then, $\log\{[lactate]/[lactic\ acid]\} = 3.5 - 3.8 = -0.30$

$$\text{antilog} - 0.30 = 0.5012$$

$$\text{i.e. } 1/2 = [base]/[acid]$$

∴ we would require twice as much lactic acid as sodium lactate. This is sensible given that the final pH required is *less* than the pK_a, i.e. the final solution is more acidic.

6. Use your answer to question 5 above and state how many grams of lactic acid and sodium lactate would be needed to prepare 2.5 litres of the buffer (pH = 3.5).

   In principle, any concentrations of lactate and lactic acid could be used providing the 1:2 *ratio* is maintained. For convenience, let us assume the total buffer concentration is to be 0.24 mol/L. To maintain the 1:2 ratio, one third of the total must be sodium lactate and therefore two-thirds must be lactic acid.

$$\tfrac{1}{3} \text{ of } 0.24 \text{ mol/L} = 0.08 \text{ mol/L}$$

   The molecular weight of sodium lactate is 112 so

$$\begin{aligned} 0.08 \text{ mol/L} &= 0.08 \times 112 \\ &= 8.96 \text{ g/L} \end{aligned}$$

   so in 2.5 litres we would need $8.96 \times 2.5 = 22.4$ g of lactate.

$$\tfrac{2}{3} \text{ of } 0.24 \text{ mol/L} = 0.16 \text{ mol/L.}$$

   The molecular weight of lactic acid = 90 so

$$\begin{aligned} 0.16 \text{ mol/L} &= 0.16 \times 90 \\ &= 14.4 \text{ g/L} \end{aligned}$$

   so in 2.5 litres we would need $2.5 \times 14.4 = 36$ g of lactic acid.

7. Refer to the diagram of citric acid shown in Section 1.v above. Draw a chemical equation to show the successive dissociation of the three ionisable protons. (*Note*: The COOH attached to the central carbon has the highest pK_a.)

$$\begin{array}{ccccccc} CH_2.COOH & & CH_2.COO^- & & CH_2.COO^- & & CH_2.COO^- \\ | & & | & & | & & | \\ OH.CH.COOH & \rightarrow & OH.CH.COOH & \rightarrow & OH.CH.COOH & \rightarrow & OH.CH.COO^- \\ | & & | & & | & & | \\ CH_2.COOH & & CH_2.COOH & & CH_2.COO^- & & CH_2.COO^- \\ & & + H^+ & & + H^+ & & + H^+ \end{array}$$

8. Figure 1.5 shows the ionic behaviour of alanine where the side chain R group is $-CH_3$. Given that the side chain for lysine is $-CH_2-CH_2-CH_2-CH_2-NH_2$, draw chemical structures showing the different ionic forms.

fully protonated form	net cation	zwitterion	fully deprotonated (anion) form
$\overset{+}{N}H_3$–$(CH_2)_4$–C(H)($\overset{+}{H_3N}$)–COOH	$\overset{+}{N}H_3$–$(CH_2)_4$–C(H)($\overset{+}{H_3N}$)–COO^-	$\overset{+}{N}H_3$–$(CH_2)_4$–C(H)(H_2N)–COO^-	NH_2–$(CH_2)_4$–C(H)(H_2N)–COO^-

9. Complete the following table by finding values where '??' is shown.

	Conjugate weak acid concentration (mol/L)	Base concentration (mol/L)	pK_a of conjugate weak acid	Final pH of the buffer solution
i	0.05	0.05	4.8	4.8
ii	0.02	0.16	3.5	4.4
iii	0.24	0.03	3.5	2.6
iv	25×10^{-3}	1.2×10^{-2}	5.1	4.78
v	1×10^{-3}	2×10^{-3}	6.5	6.8

Use the Henderson-Hasselbalch equation:

i. $pH = pK_a$, by definition when the ratio of base:acid = 1 (at 50% dissociation). Thus when acid:base ratio = 1, $pH = pK_a$

ii. Calculate the base:acid ratio first:

$$0.16/0.02 = 8$$
$$\log 8 = 0.9030$$

substitute into Henderson-Hasselbalch eqn:

$$3.5 + 0.90 = \mathbf{4.4}$$

iii. base:acid ratio

$$0.03/0.24 = 1/8(0.125)$$
$$\log 0.125 = -0.9030$$

substituting:

$$3.5 + -0.90 = \mathbf{2.6}$$

iv. base:acid ratio

$$1.2 \times 10^{-2}/25 \times 10^{-3} = 0.48$$
$$\log 0.48 = -0.3188$$

substituting:

$$5.1 + -0.3188 = \mathbf{4.78}$$

v. base:acid ratio

$$2 \times 10^{-3}/1 \times 10^{-3} = 2$$
$$\log 2 = 0.3010$$

substituting:

$$6.8 = pK_a + 0.3010$$
$$pK_a = 6.8 - 0.3010 = \mathbf{6.5}$$

10. What would be the base:acid ratio in a buffer if the final pH of the solution is 7.40 when the pK_a is 6.1? Would you expect this buffer to be effective? Give reasons.

$$pH = pK_a + \log\{[\text{base}]/[\text{acid}]\}$$
$$7.40 = 6.1 + \log\{[\text{base}]/[\text{acid}]\}$$
$$7.40 - 6.1 = 1.3 = \log\{[\text{base}]/[\text{acid}]\}$$
$$\text{antilog } 1.3 = 19.95\ (\sim 20:1 \text{ base}:\text{acid ratio}).$$

We would *not* expect this buffer to be very effective because the base:acid ratio is far from 1:1 (the ideal when $pH \approx pK_a$). However, just such a buffer system is keeping your blood pH normal even as you read this text. Details will be given in Part 3.

11. Given that pKa_2 for phosphoric acid ($H_2PO_4^- \rightleftharpoons HPO_4^{2-} + H^+$) is 6.8, calculate the base:acid ratio at normal blood pH of

7.40 (actual concentration values for $H_2PO_4^-$ and HPO_4^{2-} are not required).

$$7.40 = 6.8 + \log\{[\text{base}]/[\text{acid}]\}$$
$$7.40 - 6.8 = \log\{[\text{base}]/[\text{acid}]\}$$
$$0.80 = \log\{[\text{base}]/[\text{acid}]\}$$
$$\text{antilog}\,0.80 = \text{antilog}\,\{[\text{base}]/[\text{acid}]\}$$
$$3.98 = [\text{base}]:[\text{acid}] \sim 4$$

i.e. at normal blood pH, there is ~4× more base (HPO_4^{2-}) than acid ($H_2PO_4^-$).

Were your answers correct? If so, return to Section 1.viii and continue.

If some of your answers were incorrect, re-read Sections 1.v and 1.vi before continuing.

Self-assessment exercise 1.3

1. What is the volume (in litres) of each of the following?
   (i) total body water
   (ii) ECF
   (iii) ICF
   (iv) interstitial fluid
   in (a) a typical 70 kg man
   and (b) an 85 kg obese female (assume 1 litre of water weighs 1 kg).
   NB: you will need to make some reasonable assumptions to answer this question.
   (a) for a 70 kg adult male
      (i) Total body water (TBW) is, say, 65% of body weight = 70 kg × 0.65 = 45.5 litres
      (ii) ECF is $^1/_3$ of TBW = 45.5 × $^1/_3$ = ~15 litres
      (iii) ICF is $^2/_3$ of TBW = 45.5 × $^2/_3$ = ~30 litres
      (iv) ISF and plasma make up the ECF. We know (from Figure 1.6) that plasma volume ~3.5 litres, so 15 − 3.5 = 11.5 litres.
   (b) for an 85 kg adult female
   Females have a lower TBW than men of the same weight, plus this subject is obese so we need to estimate her TBW; let us,

for the sake of argument, agree on a figure of 55% of body weight.

(i) 85 kg × 0.55 = 46.8 litres
(ii) $^1/_3$ of 46.8 = 15.6 litres
(iii) $^2/_3$ of 46.8 = 31.2 litres
(iv) 15.6 − 3.5 = 12.1 litres

2. What is the total amount (in mmol) of (i) Na^+ and (ii) K^+ in body fluids of a 70 kg adult male?

(i) ECF volume = 15 litres; each litre contains 140 mmol of Na (see Table 1.3). ICF volume = 30 litres; each litre contains 12 mmol of Na.

$$\begin{aligned}\text{Total Na} &= (15 \times 140) + (30 \times 12)\\ &= 2100 + 360\\ &= 2460 \text{ mmol Na (2.4 moles)}\end{aligned}$$

(ii) ECF volume = 15 litres; each litre contains 4.0 mmol K.

ICF volume = 30 litres; each litre contains 148 mmol K.

$$\begin{aligned}\text{Total K} &= (15 \times 4.0) + (30 \times 140)\\ &= 4260 \text{ mmol K (4.26 moles)}\end{aligned}$$

3. Estimate the total number (amount) of osmotically active solutes in (i) ECF and (ii) ICF.

Assuming (i) a normal osmolality of 290 mmol/kg for both ECF and ICF and (ii) that 1 litre of water weighs 1 kg, we can estimate that in ECF (volume = 14 litres) and ICF (volume = 28 litres) there must be respectively 4060 mmol and 8120 mmol of osmotically active solute present.

4. Two solutions, A and B, are separated by a selectively permeable membrane. A and B are iso-osmotic and isotonic with respect to each other and both contain a mixture of large (non-diffusible) and small (diffusible) solutes and equal concentrations of anions and cations. Assume that A and B have the same volume, say 2 litres.

State with reasons what effect the following would have on the relative volumes and chemical compositions of A and B if:

(i) 500 mL of isotonic liquid containing only small *diffusible non-electrolytes* was added to A.

An equilibrium would be established as the diffusible solutes distribute equally into A and B. The volumes of both A and B would increase equally, i.e. 250 mL would be added to each compartment, as water will follow the solutes by osmosis. The solutions would reach an osmotic equilibrium as would the concentrations of anions and cations in the two compartments. The concentration of the constituents in A and B that were not contained in the added liquid would decrease due to a dilution effect.

(ii) 500 mL of isotonic liquid containing only small *diffusible electrolytes* (both anions and cations) was added to A.

The volumes of A and B would increase equally, creating a dilution effect. The added electrolytes would become equally distributed across the membrane.

(iii) 500 mL of isotonic liquid containing only large *non-diffusible* non-electrolytes was added to A.

As the added solute is restricted by its size to only one compartment (A), the tonicity of that compartment would rise and water would flow from B into A until equilibrium was reached. Thus, the volume of A would tend to increase and the volume of B would decrease, but there would also be some redistribution of the diffusible solutes present to re-establish equal tonicity.

(iv) 500 mL of isotonic liquid containing only large *non-diffusible* anionic electrolytes was added to A.

As for (iii) above but the added negative charges would cause a redistribution of cations between the two compartments in order to satisfy the Gibbs-Donnan equilibrium.

(v) 500 mL of isotonic liquid containing a mixture of small diffusible and large non-diffusible solutes was added to A.

The added small diffusible solutes would distribute equally into A and B, but the presence of the non-diffusible solute would cause a flow of water into A; its volume would increase at the expense of the volume of B.

(vi) 500 mL of hypertonic liquid containing only small diffusible solutes including electrolytes was added to A.

Equilibrium would be re-established as both ions and water distribute between both compartments. The total concentrations (and therefore osmotic pressures, i.e. osmolalities) of A and B would rise due to the addition of the concentrated solution.

5. So-called 'normal' physiological saline is prepared by dissolving NaCl in water at a final concentration of 0.85%.
   (i) What is the osmolality of this solution?

   Firstly, we need to calculate the molarity of the saline solution:

$$\text{Atomic weights: Na} = 23,\ \text{Cl} = 35.5.$$

$$\text{Formula weight NaCl} = 58.5$$

$$0.85\%\ \text{NaCl} = 0.85\ \text{g}/100\ \text{mL} = 8.5\ \text{g/L}$$

$$\frac{8.5}{58.5} = 0.1453\ \text{mol/L}, = 145\ \text{mmol/L}$$

   We assume complete dissociation of the NaCl (giving two particles) and that in dilute solution, the activity coefficient $\gamma = 1$, so the osmolality = 290 mmol/kg.

   (ii) What effect would there be on red cells suspended in 0.45% saline?

   In very hypotonic solution, the cells would swell and may even burst as water moves into the intracellular compartment, down the osmotic gradient towards an area of higher solute concentration.

6. Verify that the average charge carried by plasma proteins at normal pH is approximately 12^-. For simplicity consider only the major ions and assume that the charge on phosphate is 2 mEq/L and organic ions ~8 mEq/L.

   This answer is derived by difference of the charges carried by the main ions as shown in Table 1.3 and Figure 1.8.

Cations (mEq/L)		Anions (mEq/L)	
Sodium	140	Chloride	100
Potassium	4	Bicarbonate	26
Calcium	4.5	Phosphate	2
Magnesium	1.5	Sulphate	2
		Organics*	8
Totals	150		138
		Proteinate	12

*Organics will be a mixture of monovalent and multivalent anions, so 8 mEq/L is a weighted average.

7. Calculate (i) the osmolal gap and (ii) the anion gap for the following results:

Measured parameter	Subject 1	Subject 2
Na^+ mmol/L	135	141
K^+ mmol/L	3.8	4.2
Cl^- mmol/L	102	98
HCO_3^- mmol/L	20	27
Urea mmol/L	4.2	4.1
Glucose mmol/L	4.8	–
Osmolality mmol/kg	298	315

Subject 1:

calculated osmolarity $= 2 \times (\text{Na} + \text{K}) + \text{glc} + \text{urea} =$

$2(135 + 3.8) + 4.8 + 4.2 = 287\ (286.6)$ mmol/kg

osmol gap $= 298 - 287 = 11$ mmol/kg

anion gap $= (135 + 3.8) - (102 + 20) = 17\ (16.8)$ mmol/L

Subject 2:

We have to make a working assumption here that the glucose concentration is normal, otherwise the equation has two unknown factors:

calculated osmolarity $= 2 \times (\text{Na} + \text{K}) + \text{urea} + \text{normal glucose} =$

$2(141 + 4.2) + 4.1 + 5 = 300(299.5)$ mmol/kg

osmol gap $= 315 - 300 = 15$ mmol/kg

This is slightly higher than the 'normal' so the presence of other osmotically active compounds is indicated.

anion gap $= (141 + 4.2) - (98 + 27) = 20\ (20.2)$ mmol/L

8. At what temperatures would (i) a sample of urine with an osmolality of 450 mmol/kg and (ii) a plasma sample with an osmolality of 275 mmol/kg freeze?

(i) a 1 molal (1000 mmol/kg) solution freezes at -1.86°C

$\therefore \{450/1000\} - 1.86 = -0.837$°C

(ii) $\{275/1000\} \times -1.86 = -0.515$°C

Were your calculations correct?

If not, re-read Sections 1.viii to 1.xiii before continuing to Part 2.

PART 2

Fluid and electrolyte homeostasis

Overview

Part 2 of this text builds upon some of the fundamental physiochemical principles developed in Part 1 to explain normal and abnormal electrolyte homeostasis. Where appropriate, some re-statement of key points from Part 1 is made. Important aspects covered include the roles of the kidney in fluid balance, normal functions of selected minerals and their physiological regulation. Examples of the laboratory methods used for the quantification of electrolytes and interpretation of the results of such tests and some of their limitations are described and explained with reference to example data and case studies.

Essential Fluid, Electrolyte and pH Homeostasis, First Edition. Gillian Cockerill and Stephen Reed.

Normal physiology

SECTION 2.i

Fluid translocation: plasma to ISF and ISF to ICF

The human body is composed predominantly of 'compartmentalised water' containing a wide variety of solutes. Water within cells provides an appropriate environment to support the biochemical reactions, whilst water within the extracellular compartment essentially facilitates transport of nutrients to and waste products away from tissue cells. The solutes within body fluids create osmotic forces acting upon cell membranes. In order to avoid cells swelling or shrinking, those osmotic forces (which are of course functions of solute concentrations) exerting pressure on both sides of cell membranes must be balanced, whilst at the same time allowing the passage of solutes into and out of tissue cells. Maintenance of such a dynamic situation requires careful control if failure and consequent ill-health are to be avoided.

Nutrients enter the body via the gastrointestinal system and the lungs, and pass into the bloodstream for distribution to tissues for immediate use or to sites of storage. A normal blood pressure is required to ensure normal function of tissues and organs. Although there is a general appreciation that high blood pressure (hypertension) is a risk factor for a number of diseases of, for example, the vasculature and kidneys, a low blood pressure (hypotension) is also undesirable as this will result in poor tissue perfusion and therefore poor nutrient delivery to cells.

Essential Fluid, Electrolyte and pH Homeostasis, First Edition. Gillian Cockerill and Stephen Reed.

Clearly, though, tissue cells are not in direct contact with blood, but rather are bathed by interstitial fluid (ISF). There is then the problem of the nutrients moving out of the bloodstream across the vessel wall by capillary filtration, through the ISF and finally crossing the outer cell membrane to enter the cytosol. Conversely, waste chemicals must leave the cell, diffuse through the ISF and enter the bloodstream for transport to the kidneys, liver or lungs for excretion. Thus, a micro-circulation is established between plasma, ISF and ECF.

Fluid movement from plasma to ISF

Starling's hypothesis informs us that the exchanges between the plasma and the ISF are pressure-driven. Refer to Figure 2.1. Net flow of water and solutes is determined by the relative hydrostatic and protein-dependent oncotic pressures. At the narrow-lumen arterial end of the capillary where the effects of blood pressure are highest (approximately 4.7 kPa), there is a net outward pressure of approximately 1.35 kPa (~10 mmHg) forcing water into the ISF. This is because hydrostatic pressure exceeds the plasma oncotic pressure which tends to draw fluid from the ISF into the bloodstream. However, the blood hydrostatic

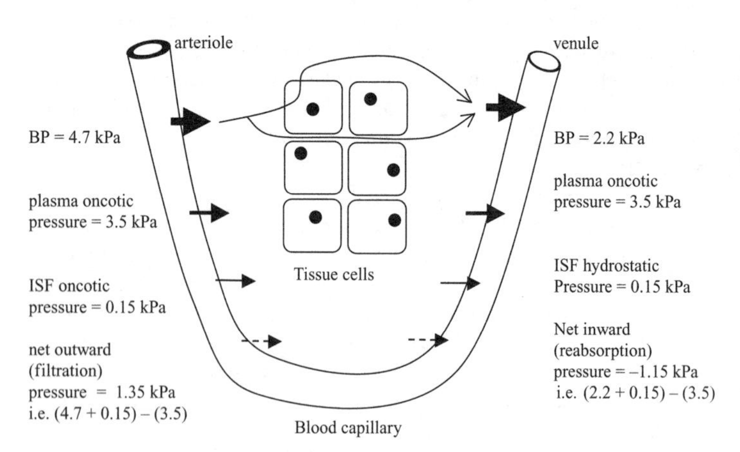

Figure 2.1 Starling's forces determine the direction of fluid movement in the micro-circulation around tissue cells
Note the pressure gradients (represented by the weight of the arrows) along the length of the arteriole and venule. Compare with Fig. 1.14.

pressure at the venous end of the capillary is less than half the value at the arterial end, and is lower than the plasma oncotic pressure (2.2 kPa compared with 4.7 kPa). Consequently, the osmotically-driven inward pressure across the vessel wall (approximately 3.5 kPa) is greater than the outward filtration pressure (1.35 kPa), and fluid returns to the bloodstream under a net pressure of −1.15 kPa (~9 mmHg). Thus, blood pressure determines the flux of fluid from plasma to ISF, whilst it is plasma protein (mainly albumin) concentration that determines fluid movement between ISF and plasma.

The volume of liquid that passes from the plasma to the ISF each day is estimated to be approximately 20 litres, considerably more than the total plasma volume of 3.5 litres, indicating significant movement out of and into the blood vascular system.

An important part of the drainage arm of this micro-circulation is the lymphatic system. Lymph is a protein-poor colourless fluid contained within thin-walled blind-ended lymphatic capillaries which lie alongside blood vessels. The walls of the lymphatic vessels are thin but have one-way valves that allow interstitial fluid to enter but not to leave. The fluid that is carried through the lymphatics eventually rejoins the blood stream via ducts which drain into the subclavian and jugular veins located in the upper part of the thorax.

Approximately 80% of filtered plasma fluid is reabsorbed directly into the bloodstream. A smaller proportion (20%) of filtered fluid and trace amounts of proteins enter the lymphatics (Figure 2.2). In certain abnormal situations the processes described above malfunction and the relative forces across the vessel wall no longer balance. If, for example, blood pressure falls (reduced hydrostatic pressure at the arterial end) or if the plasma albumin concentration is lower than normal (reduced oncotic pressure at the venous end), fluid will not return to the bloodstream normally and so the ISF volume will increase; this condition is known as oedema.

Fluid movement from ISF to ICF

The physiological regulation of osmotic balance controlling the fluid flux between ISF and ICF is regulated mainly via sodium homeostasis. Because the vascular wall is permeable to sodium ions, the [Na^+] in ISF is essentially identical to that in plasma. If ISF osmolality rises due to an increase in the [Na^+] (hypernatraemia), fluid would be drawn out of cells to equilibrate the osmotic pressures and cells would become

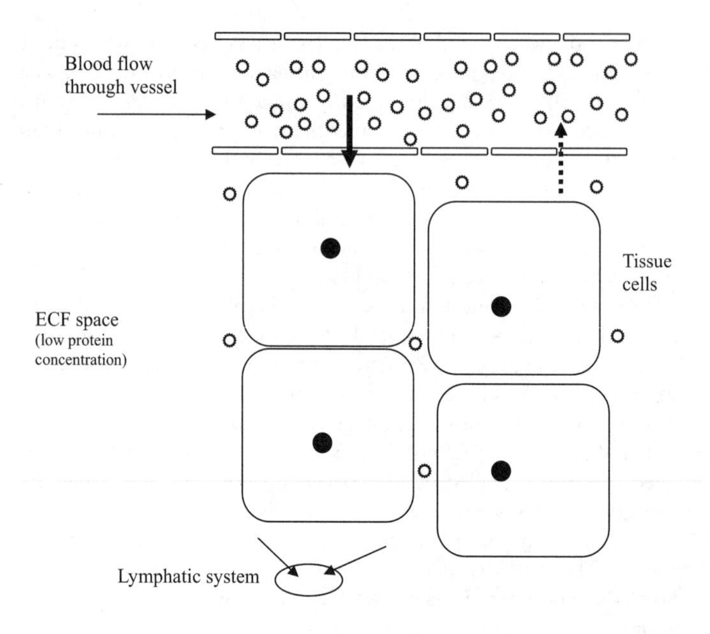

Key: → movement of water across barriers.
Weight of arrow indicates relative magnitude of the flow

Protein molecules

Blood hydrostatic pressure forces water out of the vessel shown by ↓

and the higher protein concentration in the plasma draws fluid back into the vessel ↑

Some fluid enters the lymphatic drainage system.

Figure 2.2 Drainage of interstitial space
Net fluid movement out of blood capillary at the arterial end and net fluid reabsorption into blood vessel at the venous end. Approximately 80% of filtered fluid is reabsorbed directly into the blood stream. A smaller proportion (20%) of filtered fluid and trace amounts of proteins enter the lymphatics. Protein concentration and plasma and hydrostatic pressure gradients between ISF determine water distribution across the vascular wall.

dehydrated. Conversely, a low ISF osmolality associated with hyponatraemia would cause cells to become overhydrated as fluid moves towards the area of higher osmolality within cells. Both under- or overhydration of cells leads to dysfunction and possibly overt symptoms of pathology.

Overall maintenance of fluid and water homeostasis is crucial to good health. Specific physiological mechanisms in operation are fairly simple taken in isolation, but the whole process often appears complex due to the interplay of individual organ activities and hormonal influences. In terms of volumes of fluid and quantities of electrolytes turned over each day, two organ systems that take centre stage are the kidneys and the gastrointestinal tract (Table 1.6 in Section 1.ix). In the course of a day, fully functional kidneys operating with a glomerular filtration rate (GFR) of 80–120 mL/minute will filter 140 litres of plasma water containing variable amounts of many solutes which need to be reabsorbed or excreted. Diarrhoeal diseases pose their threat to survival more through the risk of rapid dehydration than to any other cause.

The whole process of fluid and electrolyte turnover is coordinated by the endocrine and neurological systems. The key endocrine effects in the process are due to the actions of anti-diuretic hormone from the pituitary, aldosterone from the adrenal cortex, natriuretic factors from the myocardium and parathyroid hormone from the parathyroids. Neural signalling originating within the thirst centre of the brain mediates fluid intake according to the overall state of hydration by sensing plasma osmolality. The thirst response is triggered when plasma osmolality rises by even a very modest degree above the mid-range value of 290 mmol/kg.

Because of its central role in fluid and electrolyte homeostasis, and indeed acid-base balance also, it is appropriate at this stage to consider some important aspects of renal function.

SECTION 2.ii

Renal function: a brief overview

The overall structure and functions of the kidneys are probably familiar to you, so only a brief reprise will be given here. For further details, the reader should consult a standard physiology textbook such as Tortora and Derrickson (2008).[1]

Located on the rear wall of the abdominal cavity, the kidneys receive a rich blood supply amounting to 20–25% of the cardiac output per minute, which in terms of blood volume delivered per gram of tissue is proportionally greater than any other organ. Macroscopically, two distinct zones are discernable: the outer cortex and the inner medulla. Each of the two kidneys contains approximately 1,000,000 nephrons whose role is three-fold:

(i) filtration of plasma;
(ii) selective tubular reabsorption of physiologically essential components including water, minerals, glucose and amino acids; and
(iii) tubular secretion of unwanted solutes.

Nephrons consist of five anatomical regions (Figure 2.3): Bowman's capsule (which partially envelopes the knot of blood capillaries called the glomerulus), proximal convoluted tubule (PCT), loop of Henle, distal convoluted tubule (DCT), and the collecting duct. The PCT is responsible for bulk reabsorption of water and sodium in particular, whilst the DCT mediates a 'fine tuning' mechanism which carefully regulates homeostasis of Na^+, K^+, Ca^{2+}, $PO_4{}^{2-}$, H^+, $HCO_3{}^-$ and

[1] Tortora, G.J. and Derrickson, B.H. (2008) *Principles of Anatomy and Physiology* (12th edn). Chichester: John Wiley & Sons, Inc.

Essential Fluid, Electrolyte and pH Homeostasis, First Edition. Gillian Cockerill and Stephen Reed.

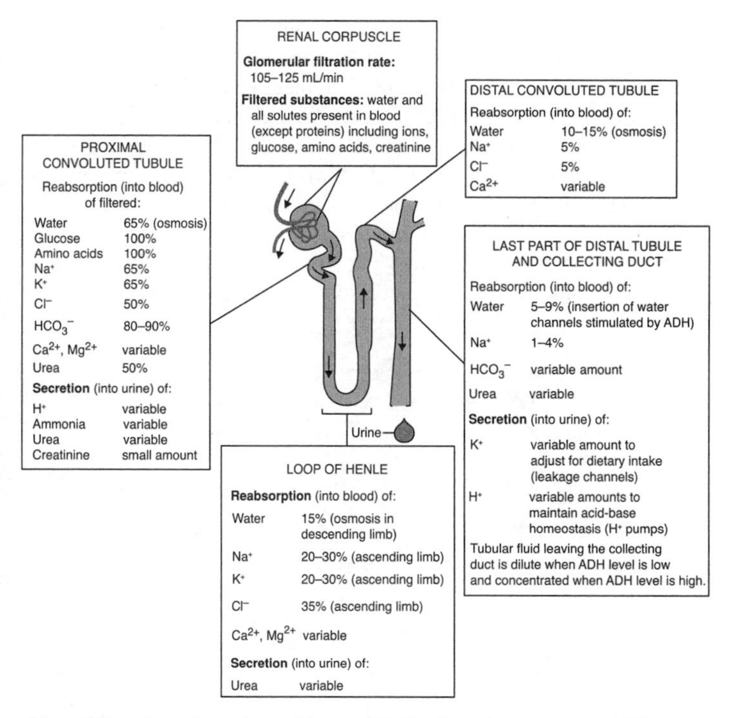

Figure 2.3 Renal nephron Figure 21.7 taken from Intro to Human Physiology by Tortora and Grabowski 6th Edition pub Wiley 2004 Figure reproduced with permission.

water. Further, the kidney can reabsorb more, or less, HCO_3^- according to the acid-base status of the ECF. This process is described more fully in Section 3.vii of this text.

The loop of Henle creates, by a counter-current mechanism, steep concentration gradients that facilitate the reabsorption of many constituents including water from the medullary region of DCT and the collecting duct also located in the medulla.

Translocation of solutes between glomerular filtrate and the renal tubular cells is achieved via a combination of diffusion and carrier-mediated mechanisms, which may be active or passive processes. Diffusion may be mediated via pores, channels, or may occur by simple permeation of the cell membrane; carrier-mediated systems require a specific membrane protein (in effect translocase enzymes) to move chemicals between the ICF and glomerular filtrate. By definition, active mechanisms are directly coupled to metabolic energy (ATP) consumption to drive the protein pumps, such as the Na/K exchanger also known as the 'sodium pump' (Figure 1.12), whilst passive transport occurs in response to a physicochemical gradient based on concentration or osmolality. On occasion, what appears at first sight to be a passive process is in fact an 'indirect' active mechanism; a gradient created by the action of an active process is exploited to bring about secondary transfer. For example, in the kidney, sodium is reabsorbed

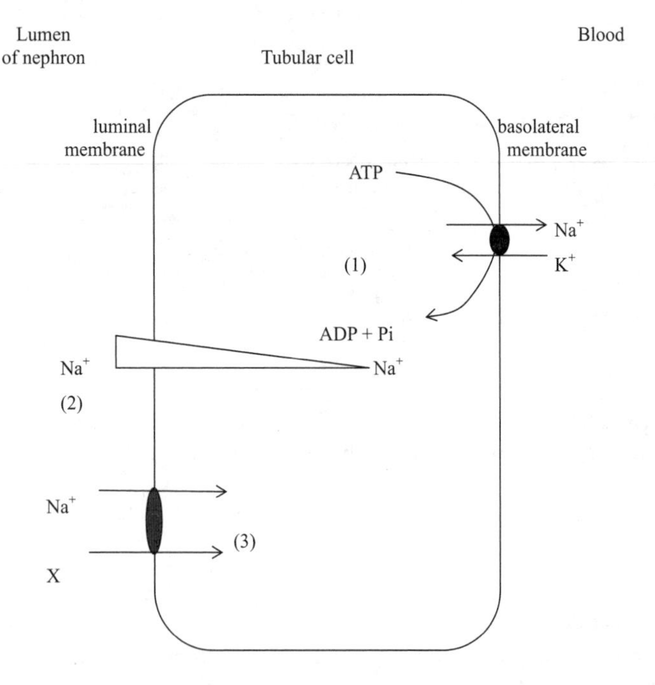

(1) Na^+/K^+-ATP'ase extrudes sodium at the basolateral membrane, reducing intracellular $[Na^+]$.
(2) Na^+ gradient is established; higher $[Na^+]$ in lumen than inside cell.
(3) Na^+ co-transporter (symport) allows uptake of X (e.g. amino acid or glucose).

Figure 2.4 Sodium gradient. Reproduced with permission from Reed, S. (2009) *Essential Physiological Biochemistry* (Wiley-Blackwell).

from the glomerular filtrate at the nephron luminal face of the proximal convoluted cells; at the same time, protons are extruded from the proximal cells. The sodium moves down a concentration gradient which is generated by the active Na/K exchanger located in the basolateral (plasma face) of the same proximal cell. Overall passive Na reabsorption at one side of the cell is dependent upon active Na pumping at the other side (Figure 2.4).

Carrier-mediated processes have a finite capacity to transfer solutes determined by their inherent efficiency and by the number of translocase proteins expressed on the cell surface. The rate of transfer that may be brought about by the carrier is termed its T_m and the maximum rate of transfer, the T_{max}, both expressed in amount of solute transferred per unit time (e.g. mg/minute). If the concentration of solute available to the transporter in the glomerular filtrate exceeds the T_{max}, net excretion of that solute will occur (Figure 2.5). Net excretion of any particular solute is a function of the GFR and the plasma concentration of that particular solute. The plasma concentration at which the solute is first detectable in urine is termed the 'renal threshold'.

Actual rates of solute reabsorption in the nephron are adaptable and subject to control. For example, parathyroid hormone simultaneously increases the rate of calcium reabsorption and phosphate excretion

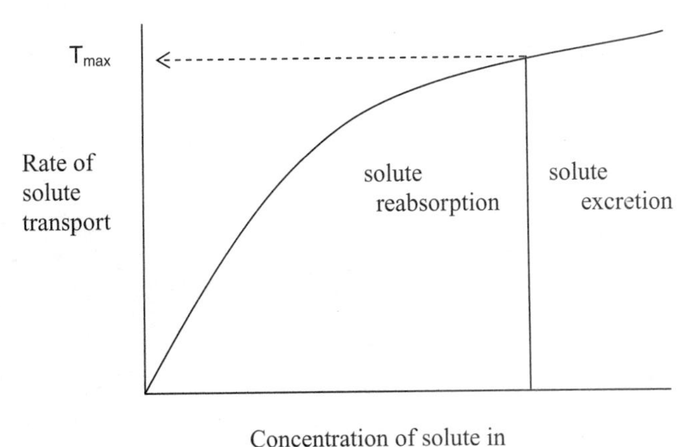

Figure 2.5 Kinetics of renal tubular transport: T_{max}

by the kidney, whereas anti-diuretic hormone upregulates (increases) the number of aquaporin water-conducting channel proteins, allowing more reabsorption of fluid from the nephron. Furthermore, a defect, usually genetically determined, in the number or activity of particular transporter proteins impairs the ability of the cell to reabsorb or secrete solutes, e.g. calcium regulation via a Ca-sensing receptor (CaSR) or indeed water as in nephrogenic diabetes insipidus.

SECTION 2.iii

Renal regulation of blood composition

In addition to its role in filtration, secretion and selective reabsorption, key features of renal physiology can be summarised as:

- Responsive to hormone stimulation;
- Synthesis of:
  1,25 vitamin D_3 (1,25 dihydroxycholecalciferol, DHCC. A sterol hormone)
  erythropoietin (peptide hormone)
  renin (enzyme)
- Metabolic roles
  degradation of small proteins and a small amount of glucose is synthesised in renal tissues by gluconeogenesis, a metabolic pathway which is especially important during periods of fasting or exercise when blood glucose concentration may fall. A key substrate for gluconeogenesis is glutamine which is also involved with proton secretion and acidification of the urine.

The kidney is the most important organ in fluid and electrolyte homeostasis. Three hormonal mechanisms operate within the kidney: anti-diuretic hormone; adrenal steroids (mainly aldosterone and to a lesser extent cortisol); and natriuretic peptides. The role of parathyroid hormone on the kidney is described in Section 2.viii.

(1) **Anti-diuretic hormone** (ADH, also called vasopressin, VP or arginine vasopressin, AVP), is a small cyclic peptide (Figure 2.6)

Essential Fluid, Electrolyte and pH Homeostasis, First Edition. Gillian Cockerill and Stephen Reed.

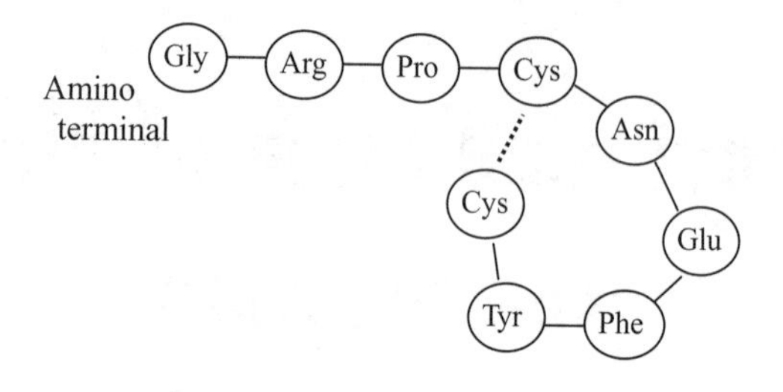

Key: Solid lines represent peptide bonds
Dotted line represents a cys-cys disulphide bridge

Figure 2.6 Amino acid sequence of anti-diuretic hormone (ADH)

synthesised by secretory neurones of the hypothalamus but released directly into the bloodstream from nerve terminals in the posterior pituitary gland. ADH/VP secretion is in response to both osmotic effects and pressure changes.

Increases in plasma osmolality above a threshold of approximately 285 mmol/kg, detected by specialised cells in the hypothalamus, trigger the release of ADH. The value of 285 mmol/kg is just within the lower end of the reference range of osmolality (284–295 mmol/kg), suggesting that there is a continuous basal secretion of ADH even under normal circumstances. Interestingly, the osmotic threshold for the thirst response is somewhat higher than the ADH threshold, but still within the normal physiological range of osmolalities, indicating that the desire to drink may be seen as a sort of 'back-up' mechanism which only occurs as plasma osmolality begins to rise. Only when plasma becomes too dilute, as indicated by an osmolality below the normal range for any reason, is ADH secretion 'switched off'. As the plasma concentration of ADH rises, the hormone begins to have an effect on the distal regions of the nephron to promote water reabsorption with maximal anti-diuretic effect being reached when plasma ADH concentration is between 3 and 4 pmol/L (see Figure 2.7). Maximal anti-diuretic effect means urine output is at a minimum and urine osmolality is at its highest, around 900 mmol/kg. When ADH concentration is very low (the actual value quoted is determined by the limit of detection of the assay used to measure the hormone), the kidney is capable of excreting more than 6 litres of very dilute urine (osmolality < 200 mmol/kg; normal value > 400 mmol/kg) in 24 hours.

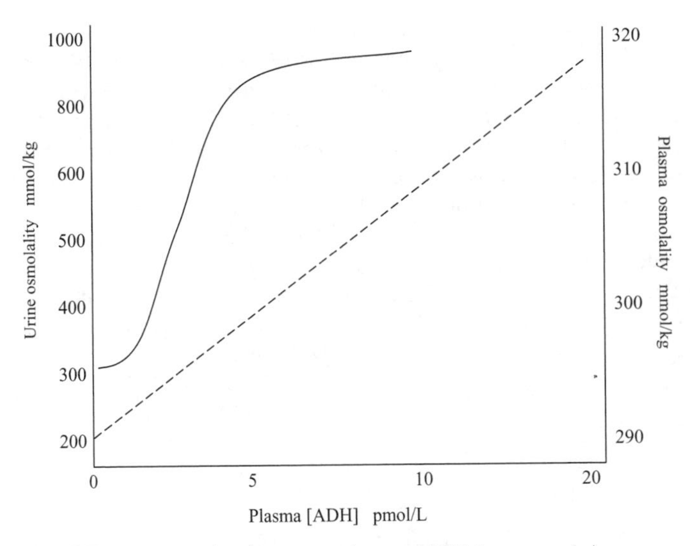

Figure 2.7 Relationship between plasma ADH (vasopressin) concentration and urine osmolality (solid line) and plasma osmolality (broken line)
Note: plotted lines represent general trends; there is wide inter-individual variation in the relationships shown.
Figure adapted and reproduced with permission from Figs 3 and 4 in Ball, S.G. (2007) *Ann. Clin. Biochem.* **44**: 417–431.

The biochemical process of water reabsorption is mediated via aquaporins (AQPs), a family of related membrane-embedded proteins which form water-conducting channels found extensively in cells of the body such as the liver, smooth muscle, central nervous system and platelets. Eleven AQPs have been described, and seven of them are to be found along the nephron; the upregulation of AQP-2 located on the basolateral face of the collecting duct is dependent upon ADH stimulation.

Physiological control of plasma volume and blood pressure is mediated by a complex neural and hormonal cascade. ADH release, part of this cascade, is itself moderated by, for example, noradrenalin (norepinephrine), which enhances ADH release whilst atrial natriuretic peptide (ANP) tends to dampen ADH secretion. If plasma ADH concentration rises to approximately 10 pmol/L, twice the upper limit of the normal range, direct vascular pressor effects become apparent. Part of this mechanism may be mediated by virtue of the fact that ADH is a weakly active corticotroph, stimulating the production of adrenocorticotrophic hormone (ACTH) from the pituitary and thus cortisol from the adrenal cortex. The major cardiovascular effect of ADH, however, is via activation of vasopressin receptors on the capillary network but only at high concentration.

Furthermore, a number of physiological 'stresses' or trauma, notably physical injury, major surgery, neuroglycopaenia (low glucose concentration in the brain, the result of hypoglycaemia) and even nausea can influence ADH secretion. A lack of ADH or the inability of the nephrons to respond to ADH stimulation gives rise to the

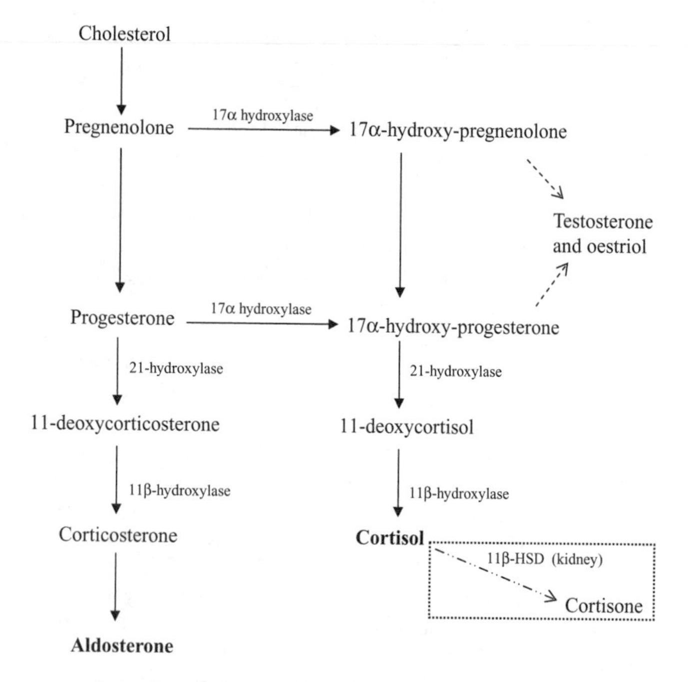

Figure 2.8 Adrenal steroid synthesis

condition known as diabetes insipidus, which loosely translated, means 'the flowing of large volumes (*diabetes*) of dilute (*insipid*) urine'.

(2) **Aldosterone**, a mineralocorticoid steroid secreted by the adrenal cortex which regulates the reabsorption of sodium and the excretion of potassium. Secretion of aldosterone is controlled mainly by the renin-angiotensin system described in more detail in the following Section. Cortisol, also an adrenal cortical hormone, has an effect on sodium reabsorption and potassium excretion in the distal renal tubule. The affinity of cortisol for the mineralocorticoid receptor is comparable to that of aldosterone, but its effect is ameliorated by the action of 11-β-hydroxy steroid dehydrogenase (11β HSD) located in the distal tubule. This enzyme modifies the structure of cortisol forming cortisone to ensure that the 'cross-reaction' between the aldosterone receptor and cortisol is minimal. However, if plasma cortisol concentration rises significantly, imbalances in Na^+ and K^+ handling in the kidney will occur as the protective action of 11β HSD is exceeded. The biosynthesis of aldosterone and cortisol from cholesterol is outlined in Figure 2.8, and their structures are shown in Figure 2.9.

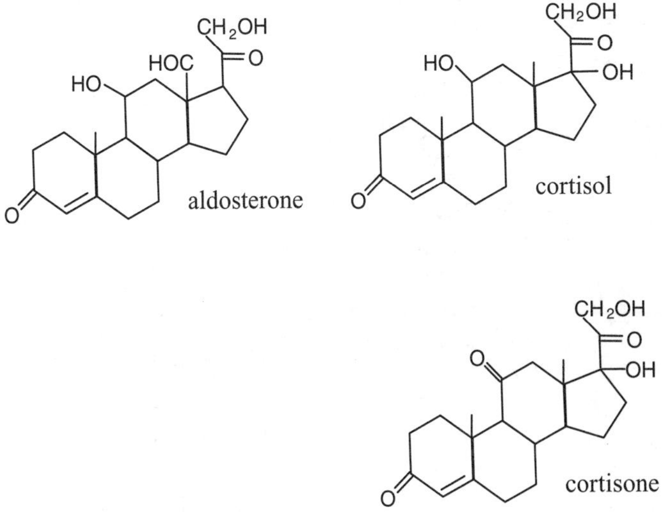

Figure 2.9 Steroid hormone structures

(3) Natriuretic peptides ANP and BNP are secreted from the heart muscle in response to stretch, an indication that the blood volume has increased. As suggested by their name, these factors promote the excretion of sodium, so oppose the action of aldosterone. The existence of a sodium-excreting hormone to avoid sodium overload and consequent potentially harmful effects of high blood pressure on the heart and vascular system had been assumed for some time before the discovery of the first natriuretic peptide (NP) in the early 1980s. Structurally, the natriuretic factors are peptides with a cysteine-cysteine disulphide bridge creating a characteristic 'loop'. All NPs have the same basic structure as shown in Figure 2.10. There are three such peptides:

- A (atrial)- type natriuretic peptide (ANP)
- B (brain)-type natriuretic peptide (BNP)
- C-type natriuretic peptide (CNP)

All three are derived from larger precursors called propeptides, each encoded by a different gene (Table 2.1).

The stimulus for secretion of the natriuretic peptides is tissue stretch, notably of the heart muscle, and responses to NPs are seen not only

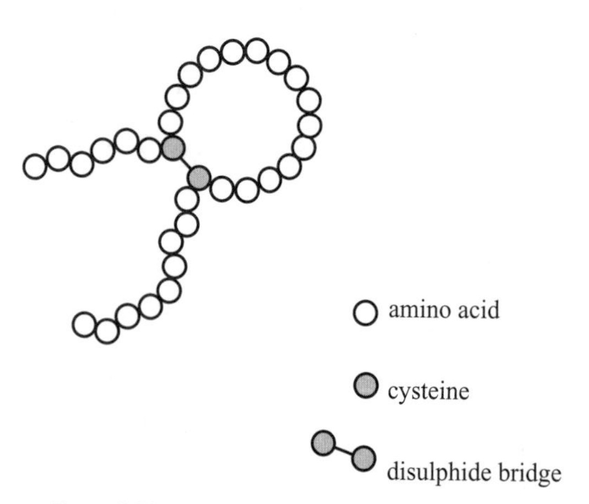

Figure 2.10 B-type natriuretic peptide (BNP)

Table 2.1 Natriuretic peptide family

Peptide	Site(s) of origin	Size of propeptide	Size of active peptide
ANP	Several tissues, but mostly from the cardiac atria	126 amino acids	28 amino acids (the C terminal of the propeptide). Other fragments of the propeptide may also have bioactivity
BNP	Brain, cardiac atria but mostly from cardiac ventricles	108 amino acids	32 amino acids
CNP	Brain, bone, vascular endothelium	103 amino acids	53 amino acids, which may in some tissues be further modified to a 22 amino acid peptide.

in the kidney (increases sodium excretion and diuresis) but also on blood vessels (causing vasorelaxation) and the adrenals.

The discussion will return to aspects of renal function at various points in the following Sections.

SECTION 2.iv

Self-assessment exercise 2.1

1. The estimated glomerular filtration rate for a subject was 90 mL/min.
   (i) What volume of glomerular filtrate is produced in 24 hours?
   (ii) If cardiac output is 5 litre/min, what volume of whole blood arrives at the kidneys each minute?
   (iii) Use your answer to (ii) above to calculate the volume of plasma which arrives at the kidneys per minute, assuming that the volume taken up by red cells, white cells and platelets is 40% of the total blood volume.
   (iv) Use your answers to (i) and (iii) above to estimate the fraction of plasma that is completely filtered at the glomeruli per minute.
   (v) Assuming each kidney contains 1 million fully functional nephrons, what volume of glomerular filtrate passes through one nephron per day?
   (vi) Assuming mid-range plasma values for [Na^+] and [K^+], calculate the quantity of each mineral filtered per 24 hours (refer to Table 1.3 for typical values).
   (vii) Refer to Tables 1.3 and 1.6. Taking mid-range values for Na^+ and K^+ excretion in urine, and using your answer to (vi) above, estimate the quantity of each reabsorbed per 24 hours.
2. Parathyroid hormone reduces the T_{max} value for phosphate. What effect would this have on phosphate excretion in the urine?

3. Some subjects who have a genetic defect causing 21-hydroxylase deficiency in the adrenal cortex have a 'salt-losing' condition. Explain why this is the case.
4. Why might a patient with *chronic* liver disease show signs of oedema?
5. What would be the physiological effect of a lack of ADH (vasopressin)?

SECTION 2.v

Minerals: key roles in physiology and metabolism

In order to maintain normal metabolism, and therefore health, the body needs a constant supply of raw materials. As well as the major food groups (carbohydrates, lipids and proteins/amino acids), water and oxygen, our diet must also contain a wide range of micronutrients which include inorganic minerals.

The physiological requirement for minerals varies considerably, and so plasma concentrations range from low micromolar to high millimolar values. From a diagnostic point of view, plasma values may not reflect total body loading, so interpretation of laboratory data can be challenging when dealing with a mineral which is mainly intracellular (notably potassium and magnesium), or one such as iron which is stored very efficiently but used in only small amounts each day. Normal plasma concentrations and total body loading of minerals are maintained by the balance achieved by dietary intake vs. losses via the gut and urinary system. Generally speaking, typical Western diets usually provide an adequate supply of most minerals and only rarely are dietary supplements required. Typical examples of the main minerals and their roles are summarised in Table 2.2 below.

As indicated in Table 2.2, hormonal influences are crucial in regulating mineral metabolism and turnover by controlling intestinal absorption or renal excretion. Furthermore, mammalian physiological systems are often very frugal and the daily turnover of a mineral may be maintained by effective recycling, that is to say that when a senescent

Essential Fluid, Electrolyte and pH Homeostasis, First Edition. Gillian Cockerill and Stephen Reed.

Table 2.2 Some important minerals

Mineral	Designation	Total amount in the body	Principal biological roles	Regulatory mechanisms
Calcium	Major metal	30 mols ~1.2 kg	Bone structure; enzyme activator; intracellular signal transduction	Hormonal; control of both intestinal absorption and renal loss
Cobalt	Minor trace metal*	1–2 mg	Component of vitamin B_{12}	Intestinal uptake
Copper	Minor trace metal	80–140 mg	Anti-oxidant; cofactor for many oxidase, cytochrome and hydroxylase enzymes	Mainly by excretion from the liver in bile via the gut
Iodine	Non-metal	30–50 mg	Formation of thyroid hormones	Intestinal uptake
Iron	Major trace metal*	4–5 g approximately 70% of this is found in haemoglobin	Redox reactions; electron transfer and oxygen binding	Intestinal uptake (duodenum) Fe (II) absorption is greater than Fe (III) at pH < 6
Magnesium	Major metal	1 mol 25 g	Enzyme activator, especially of kinases	Primary regulation unknown but a role for PTH is widely assumed.

Manganese	Trace metal	10–15 mg	Enzyme activation; antioxidant; involved with wound healing	Primary regulation unknown, probably via urinary excretion of excess dietary Mn
Phosphorus	Non-metal	25 mol ~750 g	Bone structure; metabolic organic phosphates such as ATP	Mainly hormonal actions in kidney
Potassium	Major metal	3–4 mol ~150 g	Membrane potential, neuromuscular excitability	Hormonal via control of renal loss
Selenium	Trace metal	10 mg	Component of anti-oxidant enzymes	? none
Sodium	Major metal	3.0–4.0 mols ~80 g	Osmo-regulation, cell membrane potential in concert with K, bone formation	Hormonal via renal reabsorption and loss
Zinc	Trace metal	45 mmol ~3 g	Enzyme cofactor, e.g. carbonic anhydrase and alcohol dehydrogenase	Intestinal uptake

* The terms ‘minor’ and ‘major’ to qualify trace metals relate to the total body quantity, and should not be taken to imply minor or major biochemical importance respectively.

cell is replaced by a new cell, salvage mechanisms ensure that essential minerals are not lost from the body but made available for re-use.

In addition to those essential minerals listed in the table we can add strontium, fluorine, tin, vanadium, molybdenum and chromium. On the other hand, metals such as mercury, cadmium and lead are known for their toxicity, having the ability to form chemical adducts, often via thiol (–SH) groups with, and thereby inhibiting, key enzymes. Lead, for example, diminishes haem synthesis by inhibition of two key enzymes: (i) amino levulinic acid (ALA) dehydratase, which occurs near the start of the pathway, and (ii) ferrochelatase, the final enzyme in the process, responsible for the incorporation of iron into the protoporphyrin ring.

SECTION 2.vi

Sodium and potassium

Sodium

Sodium is the principal osmo-regulator of the ECF and thus has a major impact on water distribution between ISF and ICF. The distribution of water across selectively permeable cell membranes and thus the relative volumes of the intracellular fluid (ICF) and interstitial fluid (ISF, part of the extracellular compartment) is determined principally by the [Na^+] gradient. A pathological imbalance in the distribution of sodium has an impact on fluid volumes of the ECF and ICF which may result in either (i) cellular overhydration, whose spectrum of signs and symptoms include drowsiness, confusion, coma and death, or (ii) cellular dehydration whose signs and symptoms also include drowsiness, confusion, coma and death, thus the measurement of plasma [Na^+], either at the bedside (point-of-care testing) or within a central laboratory is of critical clinical importance. Of the total body sodium, approximately 50% is found in the ECF, 10% in the ICF and the remaining 40% is termed 'exchangeable' as it is located in bone where it forms part of the mineral conferring strength upon the skeleton.

Diets often contain large quantities of sodium due to its use as a flavour enhancer or preservative, either as NaCl or monosodium glutamate, MSG. Regulation of plasma and thereby ECF [Na^+] is almost entirely dependent upon renal function. Although some Na^+ reabsorption occurs throughout the length of the nephron, the bulk occurs in the proximal convoluted tubule (PCT) and the thick ascending limb of the loop of Henle. Fine-tuning of sodium reabsorption to meet fluctuating physiological requirements is achieved in the distal tubule and collecting duct under the influence of aldosterone.

Two mechanisms operate in the PCT:

(i) an active carrier-mediated transcellular transport process which is responsive to activation of α-receptors of sympathetic nervous system, insulin and the rate of sodium delivery to the PCT, itself a function of the glomerular filtration rate and the plasma [Na]. The transporter protein carries one Na^+, one K^+ and 2 Cl^- ions simultaneously from the lumen of the nephron to the tubular cell, and is therefore called NKCC; and

(ii) passive paracellular transport in which Na moves through junctions between adjacent cells.

Sodium chloride reabsorption in the loop is carrier-mediated and in addition to physiological regulation by (for example) NaCl load arriving in the loop, sympathetic nerve activity and hormonal influences such as ADH and prostaglandin, the transport may be inhibited by so-called 'loop diuretics' such as furosemide. Drugs such as furosemide are often used to treat oedema, although they do have some unwanted side-effects on potassium, calcium and magnesium reabsorption in the nephron.

Potassium

Potassium is the major cation of the intracellular fluid (98% of total body K^+ is within cells; Table 1.3). Most of the ~75 mmol of potassium we take in each day is in the cells of the tissues we eat. The main function of potassium is to maintain electrical neutrality across cell membranes by 'balancing' the positive charge of sodium in the ECF. In particular, the concentration of plasma potassium has a profound effect on the neuromuscular junction. An acute rise (hyperkalaemia) or fall (hypokalaemia) in plasma $[K^+]$ can cause muscle weakness, paralysis and cardiac arrhythmias, possibly leading to cardiac arrest.

Plasma potassium concentration is regulated by (i) the distribution of the ion between the ECF and the ICF, which is influenced by insulin, adrenaline and the action of the Na^+-K^+ ATP'ase, and (ii) by renal excretion.

Compartmentalisation of K^+ to ICF or ECF is affected by:

(i) the action of the Na^+-K^+ dependent ATP'ase outlined in Section 1.xii. This enzyme is adaptive and responds to rises in extracellular $[K^+]$ or increased intracellular $[Na^+]$;
(ii) an increase in ECF acidity which drives protons (H^+) into cells in exchange for K^+; a fall in pH of 0.1 unit elevates plasma $[K^+]$ by approximately 0.5 mmol/L. Acidosis also causes changes in renal

excretion (see below). Conversely, a rise in ECF pH causes K^+ to enter cells as H^+ moves from ICF to ECF.

(iii) insulin, which promotes transport of K^+ into cells; dangerously high plasma concentrations of K^+ are treated by injecting insulin, and glucose, into the patient.

Renal regulation is achieved primarily by aldosterone which promotes exchange of Na^+ for K^+ (via the Na^+-K^+ ATP'ase) causing loss of K^+ in the urine, mostly via the distal tubule. Renal tubular secretion of K^+ occurs 'in competition' with secretion of protons, so potassium handling is strongly influenced by overall acid-base status. Indeed, changes in potassium or proton concentration are often linked together:

- Hypokalaemia is associated with (contributes to, or is caused by) alkalosis (high blood pH).
- Hyperkalaemia is associated with (contributes to, or is caused by) acidosis (low blood pH).

Patients with adrenocortical failure may exhibit hyperkalaemia and hyponatraemia as the lack of aldosterone allows sodium excretion and potassium retention. In situations in which there is increased urinary flow rate due to a diuretic effect within the distal renal tubule, there tends to be increased loss of K^+ in the urine. This has been termed a 'washout effect'.

SECTION 2.vii

Sodium and water homeostasis: renal regulation of blood pressure and blood volume

Maintenance of normal blood pressure is vital to the function of all tissues and organs. The heart and cardiovascular system affect blood pressure via the pumping action and vascular resistance respectively, but it is the kidneys that ultimately regulate the volume of fluid conserved or excreted in health and disease.

Hypertension can have a significant impact on, in particular, the kidneys and the central nervous system, whilst hypotension results in poor perfusion of tissues with blood and so compromised oxygen and nutrient delivery. Because the systemic blood vessels form a 'closed system', changes in volume are directly related to changes in pressure.

Blood pressure is monitored by a group of specialised cells within the kidney called the juxtaglomerular apparatus (JGA cells), and located adjacent to each Bowman's capsule act as pressure sensors. Precise control of renin release is complex, involving sensing of blood pressure in the small afferent capillary bringing blood to the glomerulus, and

Essential Fluid, Electrolyte and pH Homeostasis, First Edition. Gillian Cockerill and Stephen Reed.

the NaCl content of the fluid in the proximal tubule, which may itself involve both adenosine and prostaglandins as local paracrine signals. If blood pressure falls, the JGA cells secrete an enzyme called renin (not to be confused with rennin, an enzyme secreted by the gut). The natural substrate for renin is angiotensinogen (*angio* = vessel, *tensin* = pressure, *ogen* = precursor protein), a large plasma protein synthesised in the liver. Renin initiates a sequence of events shown in Figure 2.11. Angiotensinogen is cleaved in two steps to produce angiotensin II

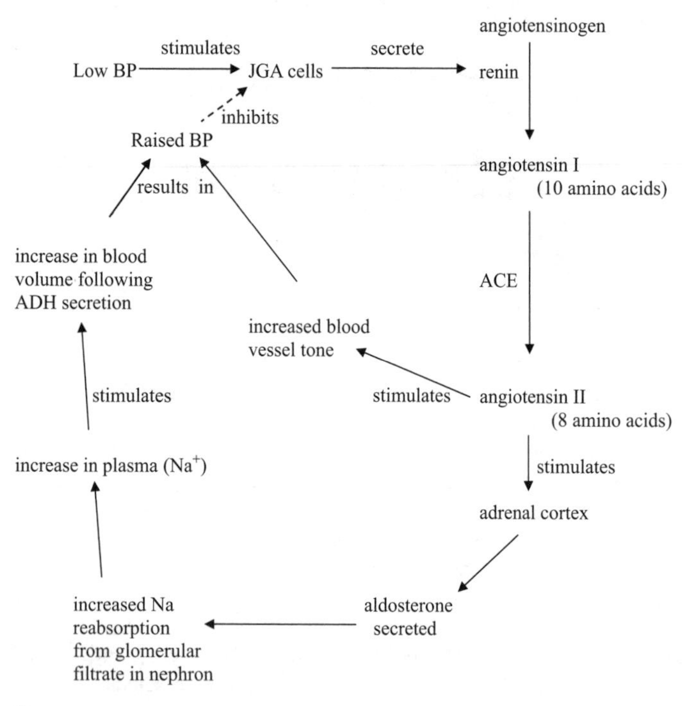

Figure 2.11 The renin-angiotensin-aldosterone (RAA) system

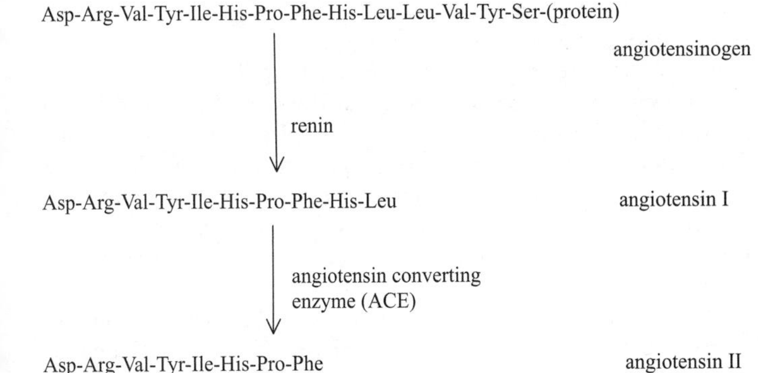

Figure 2.12 The enzymes renin and ACE produce angiotensin II

(Figure 2.12), a peptide which has four main effects, all mediated via G-protein-linked AT_1 receptors in the respective target tissues:

(i) it has vasoconstrictor actions causing increased peripheral vascular resistance;
(ii) promotion of NaCl reabsorption in the proximal convoluted tubule;
(iii) stimulation of the thirst axis originating in the hypothalamus; and
(iv) it stimulates the adrenal cortex to secrete aldosterone.

Aldosterone (a steroid hormone [see Figure 2.9, page 97]) promotes Na^+ reabsorption in the renal nephron, which in turn stimulates the secretion of ADH. ADH reduces water loss in the urine and so increases plasma volume and blood pressure, and the rise in blood pressure 'switches off' renin secretion.

Angiotensin converting enzyme (ACE) is situated in lung endothelium, but there is also local peripheral conversion in blood vessels and tissues. Of note is the fact that ACE is able to cleave a number of peptide substrates including bradykinin, a potent vasodilator. ACE-catalysed proteolysis causes inactivation of bradykinin and so potentiates the vasoconstriction brought about by angiotensin II.

Close inspection of Figure 2.11 shows the interdependent control of Na^+ and water balance: changes in fluid content in the blood regulate aldosterone; changes in plasma sodium concentration control ADH secretion. Specific cell types responsible for the reabsorption of Na^+

(aldosterone sensitive) or water (ADH sensitive) are located in the collecting duct region of the nephron.

Aldosterone exerts its action on particular cell types, called principal cells and α-intercalated cells, located within the proximal (also known as cortical) collecting duct. As is typical of steroid hormone stimulation of target tissues, aldosterone passes through the plasma membrane of the principal cell where it engages a specific mineralocorticoid receptor. The hormone-receptor complex initiates transcription of particular genes within the nucleus, resulting in upregulation of two membrane-bound transport proteins: ENaC which locates to the luminal (also called apical) membrane and allows increased uptake of Na^+ into the principal cell, whilst additional Na^+-K^+ ATP'ase proteins located in the basolateral (plasma side) of the principal cell allow the imported Na^+ to be passed into the bloodstream. To help maintain appropriate electrical balance in the cell, K^+ channels promote the excretion of potassium into the lumen of the nephron.

Aldosterone also has an action on α-intercalated cells (α-ic), which are interspersed with principal cells in the cortical portion collecting duct, but the effect here is different as it modulates a proton-linked ATP'ase which brings about bicarbonate reabsorption via the enzyme carbonic anhydrase. Interestingly, a third type of cell is believed to exist in the collecting duct. Known as the β-intercalated cell (β-ic), it has the ability to secrete bicarbonate into the nephron. Given that urine is usually slightly acidic, the concentration of bicarbonate in 'normal' urine is very low, so the β-ic cell is believed to be inactive in conditions of normal acid-base status, but if blood pH were to rise (alkalaemia), a route exists for the excretion of bicarbonate.

Plasma sodium concentration is *reduced* if blood pressure rises. Figure 2.11 shows how the kidney responds to a low blood pressure, but if plasma sodium concentration rises steeply, blood volume will rise and so too therefore will the blood pressure. Secretion of cardiac natriuretic peptides (Section 2.iii) in response to distension of myocytes is stimulated as an 'escape' mechanism.

Abnormalities leading to disruption of the balance in the forces that are involved in Starling's hypothesis may lead to excessive accumulation of fluid in tissue spaces. To summarise Starling's hypothesis briefly, plasma water is forced out of the vasculature at the arterial end of a capillary by the net outward-acting effect of the blood pressure (hydrostatic) exceeding the net inward-acting effect of the protein-based oncotic pressure. At the venous end of the same capillary, blood

(a) Normal

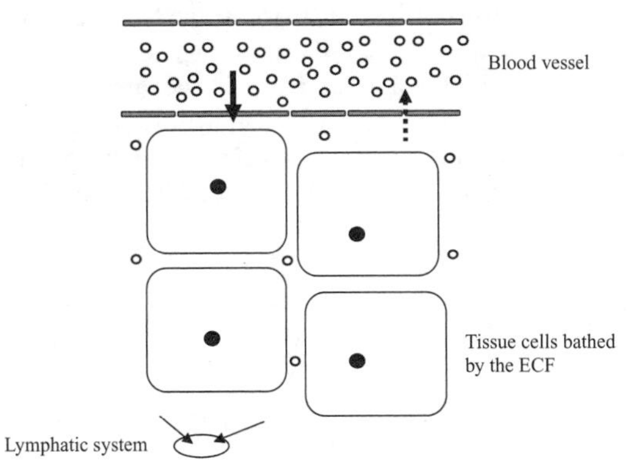

(b) Hypoproteinaemia. Plasma oncotic pressure (due to protein concentration) is lower than normal so water movement from ISF to plasma is reduced and fluid accumulates in the ECF tissue space as oedema.

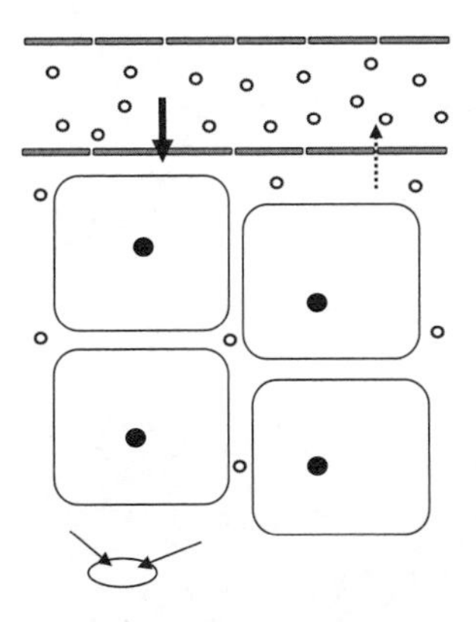

Key:
- (solid arrow) water movement across endothelial barrier
- (thin arrow) drainage into lymphatics
- (dotted arrow) fluid return to bloodstream due to plasma oncotic pressure

Figure 2.13 Oedema

pressure is lower and now the oncotic pressure exceeds hydrostatic pressure and fluid returns to the plasma space. If the balance between these forces becomes compromised, ISF volume increases and is called oedema (Figure 2.13). Fluid that accumulates in the interstitial space is termed an exudate associated with an inflammatory reaction, or a transudate arising from the leakage of protein-poor fluid from the plasma into tissue spaces. These processes are described more fully in Section 2.xx.

SECTION 2.viii

Calcium and magnesium

Calcium

The total amount of calcium in the typical adult is in excess of 30 moles, which is approximately ten-fold more than the total body content of either sodium or potassium. Not surprisingly, the vast majority (99%) of this amount is to be found in bone in the form of hydroxyapatite crystal, a composite of calcium phosphate and calcium hydroxide ($3\{Ca_3(PO_4)\}Ca(OH)_2$). Only about 7.5 mmol (≈0.025%) of the total body load of calcium is circulating in the bloodstream. These figures mask the significant turnover of calcium in the body (Figure 2.14), with large quantities being redistributed among tissues and body fluids each day. Plasma total calcium (tCa) occurs in three 'forms':

- free ionised Ca^{2+} (approximately 49% of the total);
- protein bound, mostly to albumin (48%);
- complexed with anions such as phosphate (3%).

The ionised fraction is physiologically active and in dynamic equilibrium with the protein-bound fraction which acts a 'reservoir' or buffer to maintain the correct concentration of free Ca^{2+}.

$$\text{Albumin-Ca} \rightleftharpoons \text{free } Ca^{2+}$$

Note that this equilibrium is affected by blood pH; acidosis pushes the equilibrium to the right so increasing the proportion of free Ca^{2+} and vice versa. Changes in plasma albumin concentration affect the total calcium concentration, but the position of equilibrium is maintained and so the $[Ca^{2+}]$ remains normal. Mathematical corrections of apparently abnormal results for Ca^{2+} or tCa are possible. For example,

Essential Fluid, Electrolyte and pH Homeostasis, First Edition. Gillian Cockerill and Stephen Reed.

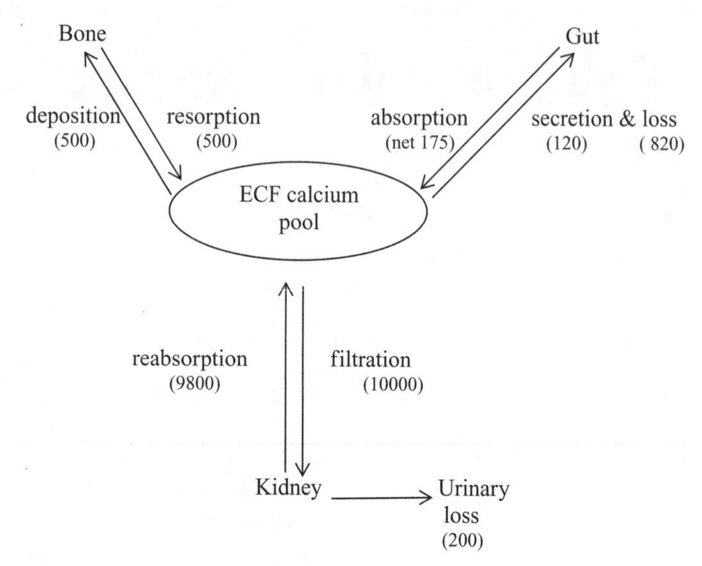

Figure 2.14 Calcium distribution and tissue turnover (values are mg/24 hours)

to correct for hypoalbuminaemia:

$$\text{Corrected [tCa]} = \{\text{measured [tCa]}\} + 0.02 \times \{41 - [\text{measured albumin}]\}$$

(41 is taken as typically 'normal' for plasma albumin concentration (41 g/L). Some equations use a value of 40 instead of 41 in this context.)

Furthermore, plasma ionised calcium concentration falls as blood pH and/or body temperature rise. A similar effect is seen with magnesium concentrations. Interpretation of laboratory results for calcium and magnesium can therefore be a little tricky. Currently, it is not routine practice in clinical laboratories to correct measured values for either ion for changes in pH or temperature.

Ionised calcium is also functionally important within cells. As shown in Table 1.3 on page 30, the ICF concentration of Ca^{2+} is approximately 4 orders of magnitude lower than the plasma concentration.

A major function of intracellular calcium in tissues other than muscle is as a message relay system, promoting changes in enzyme activity within the cell in response to a stimulus from a hormone. In muscle, a rise in cytosolic calcium ion concentration is a trigger for contraction.

A typical Western diet provides 10–20 mmol of calcium per day, most of which is absorbed in the small intestine. The actual amount of calcium absorbed each day is variable according to need, and tends to decrease with age. In health, this intake is matched by losses through the kidney (5 mmol/day) and gut (20 mmol/day). Intestinal absorption is promoted by vitamin D (see later in this Section) but impaired by fatty acids and phosphates present in the diet.

A study of the homeostatic regulation of plasma calcium concentration illustrates very well the integrated functions of several organs and hormones. Physiological maintenance of *ionised* calcium concentration at approximately 1.2 mmol/L is critical for normal neuromuscular activity. Muscle spasms or twitching (tetany) can occur if ionised calcium concentration falls significantly. The major organs involved with calcium homeostasis are the kidney, the gut, the bone and the parathyroids.

Calcium uptake in the gut, loss through the kidneys and turnover within the body are controlled by hormones, notably parathyroid hormone (PTH) and 1,25 dihydroxy cholecalciferol (1,25 DHCC or 1,25 dihydroxy vitamin D3). Other hormones such as thyroxine, sex steroids and glucocorticoids (e.g. cortisol) influence the distribution of calcium. The role of calcitonin is probably not significant in humans except in normal growth in childhood and some abnormal situations when calcium turnover is accelerated.

Parathyroid hormone is a peptide of 84 amino acids. The active hormone is synthesised as a larger peptide pre-pro-PTH (125 amino acids) which undergoes post-translational processing to pro-PTH (90 amino acids) within each of the four parathyroid glands. The main physiological action of PTH is to raise the plasma concentration of calcium by (a) increasing renal reabsorption (of calcium and magnesium), (b) increasing the synthesis of 1,25 vitamin D3 by controlling the activity of the 1α hydroxylase enzyme in the kidney (see below), and (c) promoting calcium loss from bone mineral by stimulating osteoclast cells. Other effects of PTH on the kidney include alkalinising urine by decreasing proton secretion and bicarbonate reabsorption; promoting phosphate loss through the kidney (phosphaturic effect); and decreasing sodium reabsorption. PTH secretion is controlled by negative feedback by [Ca^{2+}]. Another peptide, calcitonin, has some effects on calcium metabolism but its significance in health and disease is uncertain. Parathyroid hormone related protein (PTHrP) is a close structural homologue of PTH but is produced by tumours, especially of the lung.

Plasma calcium ion concentration is monitored by calcium-sensing receptors (CaSRs) located in the cell membrane of C cells of the parathyroid gland and of cells in the loop of Henle and the distal nephron. Structurally, CaSRs are composed of approximately 1100 amino acids with a long extracellular domain of 612 amino acids and a shorter intracellular domain (approximately 210 amino acids) which operates via a G-protein/phospholipase signalling mechanism. The transmembrane domain has the typical 7-pass 'serpentine' arrangement of many G-linked surface receptors.

Engagement of calcium receptors results in mechanisms designed to lower plasma calcium concentration. In the parathyroid gland, calcium binding to CaSR causes a reduction in secretion of parathyroid hormone (PTH), whilst CaSR activation in the nephron results in reduced calcium reabsorption from glomerular filtrate and thus increased calcium loss in urine (hypercalciuria). Genetic defects in the structure and therefore function of CaSR lead to disorders of calcium homeostasis. For example, single amino acid substitutions may result in 'loss of function' mutations causing hypercalcaemia with hypocalciuria because PTH secretion and renal reclamation of calcium are not suppressed, allowing the plasma calcium concentration to rise. Typically, plasma total calcium concentration in affected individuals is elevated, up to 3.20 mmol/L in some cases (reference range 2.25–2.55 mmol/L). Conversely, activating mutations, again usually single amino acid substitutions, bring about hypocalcaemia (total calcium < 2.20 mmol/L) with hypercalciuria, a condition which mimics parathyroid deficiency (primary hypoparathyroidism).

Vitamin D3 is a fat-soluble steroid-like molecule which elevates plasma calcium concentration. A typical diet contains some preformed vitamin D2 (from plant sources) and vitamin D3. Endogenous synthesis of dihydroxy vitamin D3 begins with 7-dehydrocholecalciferol in the skin, which, upon exposure to solar ultra-violet light (wavelength of $\sim$300 nm), is converted into vitamin D3. Two hydroxylation reactions then occur, firstly in the liver (25 hydroxylase) and then in the kidney (1α hydroxylase). Refer to Figures 2.15 and 2.16. Functional

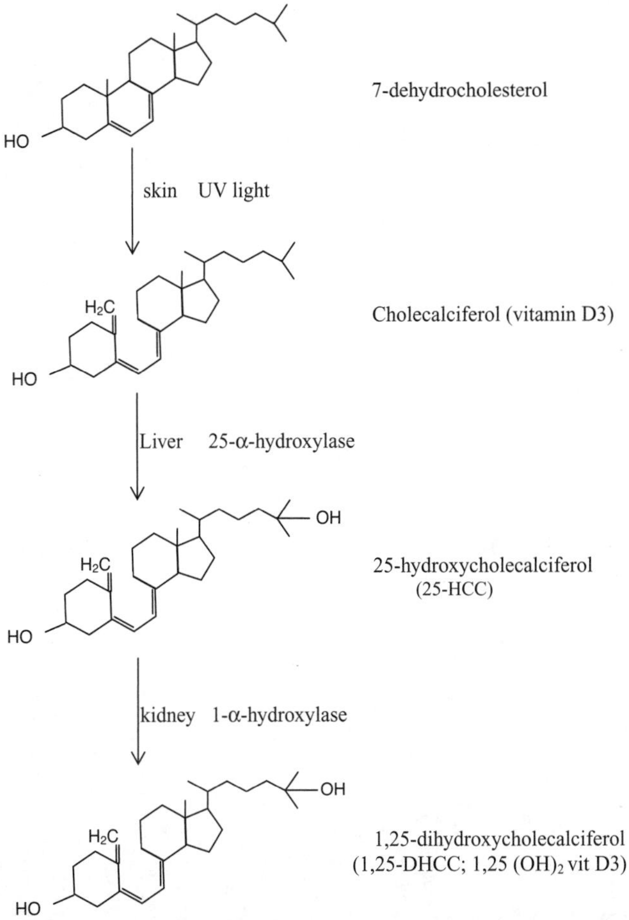

HO

7-dehydrocholesterol

skin UV light

H_2C

HO

Cholecalciferol (vitamin D3)

Liver 25-α-hydroxylase

OH

H_2C

HO

25-hydroxycholecalciferol
(25-HCC)

kidney 1-α-hydroxylase

OH

H_2C

HO

1,25-dihydroxycholecalciferol
(1,25-DHCC; 1,25 $(OH)_2$ vit D3)

Figure 2.15 1,25 dihydroxy vitamin D3 synthesis. Reproduced with permission from Reed, S. (2009) *Essential Physiological Biochemistry* (Wiley-Blackwell).

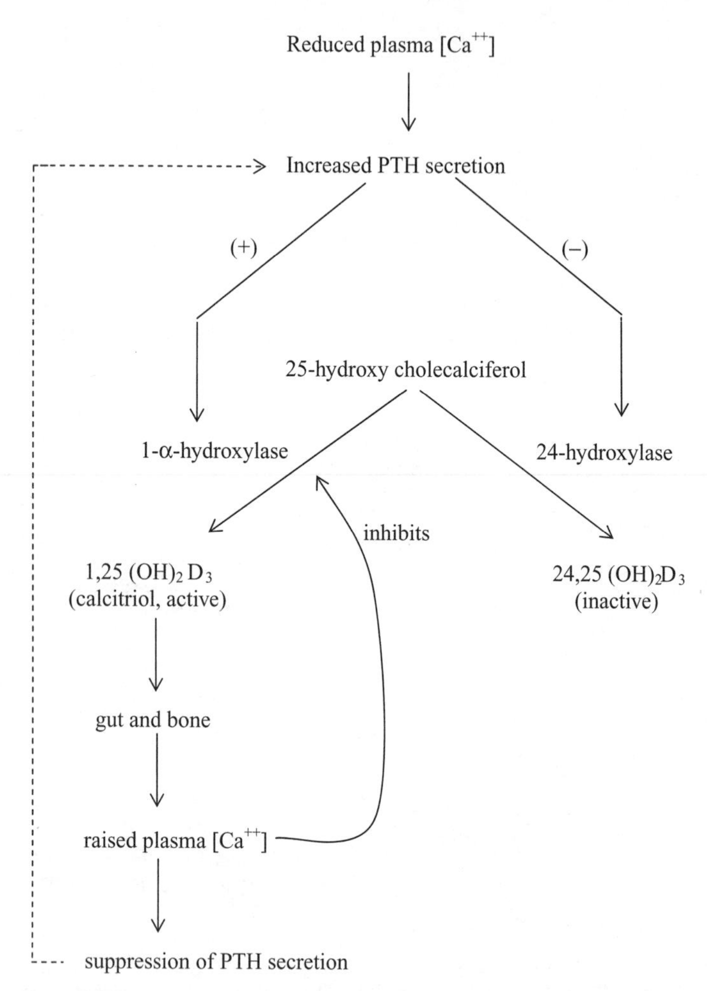

Figure 2.16 Feedback control of dihydroxy vitamin D3 synthesis by ionised calcium

1,25 dihydroxy vitamin D3 mobilises calcium from bone, and promotes calcium and phosphate uptake in both gut and kidney.

Magnesium

Magnesium is second only to potassium in terms of importance as an intracellular ion, but more than 60% of the body's total Mg (approximately 25 g, ~1 mol in an adult) is to be found associated with calcium in bone mineral. Plasma magnesium, 0.7–1.1 mmol/L, accounts for only 1% of the total. Like calcium, magnesium in plasma is found protein-bound (35%), chelated with anions (12%) and as free ionised Mg^{2+} (53%). Intracellularly, magnesium is a cofactor for numerous, possibly as many as 300, enzymes involved with metabolism of carbohydrate, protein and lipids. The Mg^{2+} ion is often complexed with ATP and is an essential cofactor for phosphorylation reactions catalysed by kinases. The Na/K ATP'ase pump (Figure 1.12) is a kinase, so the cytosolic concentration of Mg^{2+} affects sodium and potassium exchange across cell membranes and thereby contributes to normal ionic balance.

A balanced diet consisting of cereals, green and leafy vegetables, nuts, fruit, animal protein or seafood supplies the recommended minimum intake of 0.25 mmol/kg body weight each day. Magnesium is absorbed mainly in the duodenum and to a lesser extent in the large intestine. Most of the daily losses are via the urine (up to 5 mmol/day) and gut. There is no known 'key' magnesium-controlling hormone. However, magnesium uptake into cells is influenced by insulin, and, significantly, PTH may help to regulate magnesium turnover. Magnesium is a constituent of many laxatives and 'over the counter' purgative medicines, and excessive use of such preparations can lead to diarrhoea.

Abnormalities in magnesium and calcium control are often coincident because the operation of calcium channels is dependent upon Mg, and also because the release and action of PTH are influenced by magnesium. Clinically, the signs of hypomagnesaemia (<0.6 mmol/L) are identical to those of hypocalcaemia: tetany, seizures and neuromuscular irritability. Cardiac arrhythmias may be related to hypokalaemia arising from impaired action of the sodium pump as indicated above.

Like calcium, the non-protein bound fraction of total Mg is freely filtered at the glomeruli, but most of this amount is reabsorbed in the thick ascending limb of the loop of Henle. The mechanism of this reabsorption is indirectly active in that the movement is dependent upon an electrochemical gradient maintained by active pumping of other ions. Additionally, the tubular reabsorption of magnesium is inversely related to the flow rate of glomerular filtrate so patients with poor renal function (and thus a low glomerular filtration rate) tend to be hypomagnesaemic.

The relationship between magnesium and insulin is intriguing. Insulin promotes Mg movement into cells in a glucose-independent fashion, and once inside cells, magnesium is involved in mediating the actions of insulin. Consequently, a low intracellular concentration of Mg^{2+} is associated with insulin resistance. By mechanisms which are not clearly understood, magnesium regulates blood vessel tone and therefore blood pressure; a low intracellular concentration of magnesium is associated with hypertension. Furthermore, hypomagnesaemia has been implicated with an increased risk for the development of atherosclerotic vascular disease. The mechanism of this association is complex, but activation of endothelial cells that line the vasculature leads to an increase in the expression of cytokines (various growth factors in particular), and upregulation of adhesion molecules which slow and finally arrest the flow of white blood cells across the endothelium. Further evidence suggests that low cellular magnesium concentrations lead to the overproduction of reactive oxygen species (ROS), oxidation of low-density lipoprotein (LDL), and inflammation and activation of the blood coagulation cascade system. All of these phenomena are associated with the pathogenesis of atheroma. Not surprisingly, therefore, the use of magnesium infusions in the treatment of acute myocardial infarction (heart attack) has been suggested.

SECTION 2.ix

Iron

Every cell of the body contains iron, either as haemoglobin or in cytochromes found in the mitochondria and/or microsomes; catalases and peroxidases also require iron as part of a prosthetic group to perform their oxidative roles. Given the number of essential enzymes and proteins that require iron to fulfil their functions, there is surprisingly little (approximately 5 g or about 85 mmol, compare with ~3200 mmol of Na^+) of this mineral in the human body (see Table 2.2, Section 2.v). In common with many other minerals, iron exists in variable oxidation states; Fe (II) is soluble but toxic because of its capability to prompt the formation of damaging free radicals such as superoxide ($O_2{\cdot}^-$) and hydroxyl ($OH{\cdot}^-$), and Fe (III) is non-toxic but insoluble. It should be noted, therefore, that nearly all of the total body iron is in fact protein-bound (Table 2.3).

The main storage proteins for iron, ferritin and haemosiderin, are found in several tissues notably the gut, the spleen, the liver and the erythropoietic tissues. Haemosiderin has by far the larger storage capacity of the two principal proteins, and its main function appears to be to allow slow release of iron. This protein may be visualised microscopically via the Perls' Prussian blue reaction applied to sections of tissue such as liver or bone marrow aspirates. Transferrin (molecular mass ~ 450,000) consists of a shell of protein of up to 24 individual peptide subunits of H (heavy) or L (light) chains and can hold over 4000 atoms of iron, which are readily available when required. The H-type isoform has inherent ferroxidase activity and can therefore convert Fe (II) to Fe (III).

The ability of iron to bind with oxygen accounts for its role in haemoglobin (2.5–3 g in total) and myoglobin, the reddish pigment found in skeletal muscles, where it acts as an oxygen 'reserve' for periods of vigorous exertion when the metabolic demand for oxygen outstrips its delivery. The circulating plasma concentration of iron is

Essential Fluid, Electrolyte and pH Homeostasis, First Edition. Gillian Cockerill and Stephen Reed.

Table 2.3 Distribution of iron

Location	Quantity
Circulating in plasma bound to transferrin	3 mg
Liver as ferritin and haemosiderin	~1 g
Myoglobin	0.5 g
Haem-containing enzymes	400 mg
Bone marrow	300 mg
Red blood cells as haemoglobin	2.5–3 g
Macrophages	600 mg

approximately 10–30 μmol/L, but measurement of serum iron is a very poor indicator of total body iron status, partly because (as shown in Table 2.3) most is intracellular but also because there are marked diurnal fluctuations, with higher values, by as much as 30%, being found in the morning than in the afternoon and evening.

There is no controlled excretion of iron, so homeostasis is achieved via intake and tissue distribution. Typical losses amount to about 1 mg/day but this value can nearly double during heavy bleeding or menstruation. Intestinal uptake of iron from the diet is approximately 2 mg/day, yet the daily turnover is approximately 25 mg, indicating that there is efficient recycling of iron within the body. Approximately 80% of the recycled iron is from senescent ('old') red blood cells each day. All such old red cells are removed from the circulation by the reticulo-endothelial cells of the spleen and the liver, but also by macrophages, and these cells have a major role to play in recycling iron. Furthermore, the daily turnover is quantitatively much greater than the plasma iron concentration, indicating that movement into cells must be both an efficient and a well-controlled process, and one which is now known to be heavily dependent upon membrane transporters.

The diet provides iron typically around 15 mg/day usually in the form of elemental non-haem ferric (Fe (III)) iron and, in meat-eaters, as haem-bound iron. In vegetarians, most of the daily iron is derived from cereals. Although the mechanism of the intestinal absorption of haem is not well understood, it is a more efficient means of iron uptake than is absorption of non-haem iron. Processes involved with uptake of elemental iron are better described than are those for the absorption of non-haem iron. Only ferrous (Fe (II)) iron can be absorbed in the gut but because most dietary iron is in the form of Fe (III), a ferric reductase (itself an iron-containing cytochrome protein) is required for the conversion of Fe (III) to Fe (II). Translocation of Fe (II) is mediated by a divalent metal transporter-1 (DMT1) located on the luminal side of the enterocytes that line the duodenum. A number of factors affect elemental iron uptake; gastric acidity enhances uptake because it maintains iron in the Fe (II) state, whilst a diet rich in organic phosphates (e.g. phytates found especially in cereals) decreases iron absorption. Intestinal absorption of iron is adaptable to physiological demands, so there is a rise in absorption in times of overall deficiency or during periods of increased erythropoietic activity such as iron deficiency, pregnancy, haemolysis, bleeding, or at times of living at high altitude (chronic hypoxia). Once inside the cells of the gut mucosa, iron is bound to apo-ferritin to form ferritin, much of which is lost from the body via the faeces as the luminal cells are sloughed off in normal, and rapid, cell turnover in the gut.

Export of iron via the basolateral face of the gut cells is via a protein called ferroportin (Fpn). This protein, which is also found in hepatocytes, macrophages and cells of the placenta, is, as yet, the only iron efflux transporter to have been described. At the time of export from the enterocytes, Fe (II) is re-oxidised by a ferroxidase, a reaction probably involving a Cu^{2+}-containing caeruloplasmin. Transport of iron as Fe (III) through the bloodstream is mediated by the beta-globulin transferrin (Tf). Normally, circulating Tf carries iron to only about 30% of its maximum capacity, and most of this iron is derived from macrophages. A summary of the mechanisms of iron turnover is given in Figure 2.17.

In the event that particular cells, especially the haemopoietic tissues, become iron depleted, the integrated action of two iron regulatory proteins (IRPs) within the cytosol allows increased uptake of iron into the cell. IRP binds to mRNA coding for proteins which are involved with iron storage and transport, notably transferrin receptor (TfR), a membrane-bound transporter. Increased expression of TfR on the cell surface allows more Tf-Fe (II) uptake, thus improving iron availability. Other proteins similarly controlled by IRP binding to mRNA include ferroportin, ferritin and DMT1, all of which, along with Tf and TfR, are involved with iron uptake and utilisation. However, there is one other protein, discovered and isolated in 2000, which seems to have an overall controlling influence on iron distribution and homeostasis. The protein is called hepcidin, so-called because it is synthesised in the liver (*hep*) and was originally identified due to its *in vitro* bacteriostatic

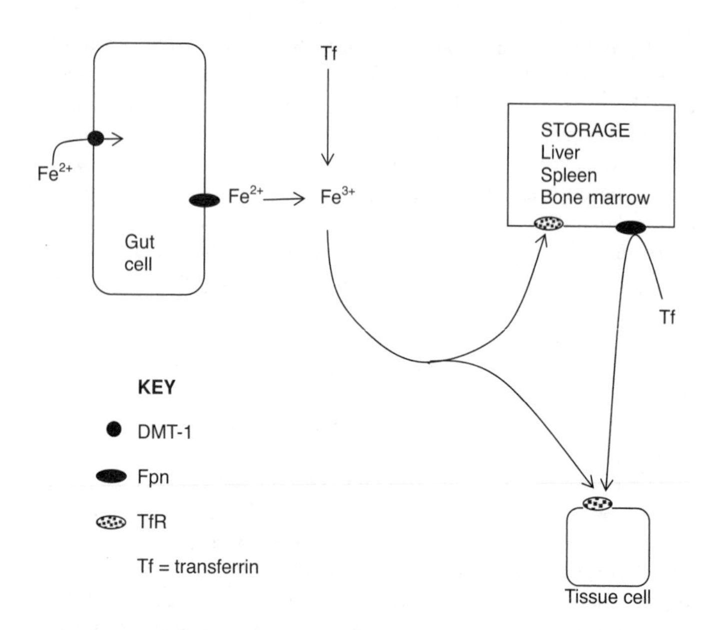

Figure 2.17 Iron uptake and distribution

properties (*cidin*). Whether hepcidin has any physiological role as an antimicrobial or antifungal is doubtful as the concentrations found *in vivo* are at least two orders of magnitude lower than those used experimentally.

Hepcidin is a small (25 amino acids) cysteine-rich peptide, which is highly conserved throughout vertebrate species, whose main function is as a negative regulator of iron uptake and tissue distribution. Experimental under-expression of hepcidin (in a mouse model) is associated with haemochromatosis, a genetic disorder in which there is inappropriate intestinal absorption and increased storage of iron; conversely, overproduction of hepcidin results in iron deficiency. It appears that translation of these findings to humans is possible. The mechanism of action of hepcidin is by inhibiting Fpn-mediated uptake of iron in the gut, placenta, macrophages and liver.

Production of hepcidin is controlled by body iron stores (most likely transferrin and/or ferritin concentrations) and erythropoietic activity in the marrow. Further, hepcidin synthesis is:

(i) *suppressed* in cases of alcoholic liver damage (possibly accounting for the haematological changes and increased iron stores often seen in alcoholics), in cases of chronic hepatitis, anaemia and hypoxaemia (low oxygen content); and
(ii) *increased* in by interlukin-6 (IL-6), a key player in the initiation of inflammation, an observation which explains why patients with chronic disease often become anaemic despite adequate dietary iron.

SECTION 2.x

Selected trace elements: Mn, Co, Se and S

Manganese is a fairly abundant metal in the environment and is an indispensable metal for the normal function of plants and animal cells. Estimated dietary intakes of at least 2 mg are more than sufficient to prevent deficiency, although intestinal uptake is only around 5% of the ingested load, with the excess easily being excreted. Intoxication leading to neurological and psychological disorders can occur especially in persons working in metal-related occupations such as mining, smelting and steel industries. The total body loading of Mn is around 15 mg and this occurs in several oxidation states, with Mn^{2+} and Mn^{3+} being the most common. Most of the circulating Mn is bound to albumin (as Mn^{2+}) with a little combined with transferrin (as Mn^{3+}); the latter should not be surprising when we consider that Fe and Mn are adjacent to each other in the periodic table. Liver, kidney, pancreas and bone are the main tissues in which Mn is to be found. In soft tissues, Mn is located mainly in the mitochondria and nuclei, often associated with a wide range of enzymes such as RNA polymerase; Mn is also is implicated in the action of the respiratory cytochromes, arginase, pyruvate carboxylase and superoxide dismutase, and also for glycosyl transferases which are important in the synthesis of extracellular matrix of connective tissue, thus establishing its role not only in normal growth but also in repair and wound healing.

Cobalt is an essential component of vitamin B12 and so is involved, often in concert with folate, with '1-carbon metabolism', i.e. the transfer of functional groups containing only one carbon atom. An important example of such a process is the synthesis of purines and pyrimidines. Estimates of daily gastrointestinal intake of Co vary from less than 1 μg to around 40 μg; a small quantity of the metal may

Essential Fluid, Electrolyte and pH Homeostasis, First Edition. Gillian Cockerill and Stephen Reed.

also be absorbed via the respiratory system and this route represents a major risk of occupational toxicity in workers in the metal industries. Approximately 10% of the total body Co is found in the liver, and excretion is via both urine and gut.

Selenium is a non-metal trace element whose main physiological function is associated with antioxidant enzymes such as superoxide dismutase (SOD), as is manganese. Se-SOD is one of the body's defence mechanisms against damage arising from the generation of free radicals as a result of normal metabolism, the metabolism of xenobiotics, and as part of the immune response to infection. Total body load is approximately 10 mg and the recommended daily intake is around 60 μg. There seems to be no physiological control over selenium uptake in the small intestine, although absorption is often in association with a sulphur-containing amino acid such as cysteine or the tripeptide glutathione. Excessive dietary loads, possibly from an over-zealous use of vitamin and mineral supplements, lead to selenosis causing hair loss, brittle nails and some neurological symptoms such as numbness.

Sulphur is a non-metal which is present in such large quantities in the body that it does not really qualify as a trace element. The main interest in sulphur in the context of fluids and electrolytes is as a source of metabolic acid, because its oxidation leads to the formation of sulphuric acid, albeit in fairly small quantities each day. A typical diet, and especially one rich in meat protein, will contain more than adequate quantities of sulphur in the form of sulphur-containing amino acids and as sulphite, a food preservative. Deficiency states are essentially unknown and circumstances leading to overload are very rare. From a functional point of view, sulphur occurs as sulphate in complex biomolecules such as glycosaminoglycans (heparin sulphate, chondroitin sulphate) which are themselves important constituents of connective tissues. Sulphydryl (also called thiol) –SH groups are significant in the action of many enzymes.

SECTION 2.xi

Anions: bicarbonate, chloride, phosphate and proteins

Chloride is the principal anion of ECF where it serves as a major counter-ion to sodium. Because of this close association with sodium, chloride balance is maintained mainly via renal tubular reabsorption. In the proximal convoluted tubule, chloride transport is 'passive', being secondary to reabsorption of sodium, but the loop of Henle has an active chloride transport mechanism which helps to maintain the locally high osmotic pressure of the cortical tissue required to ensure water reabsorption via the counter-current effect.

Quantitatively, bicarbonate is the second most important anion in ECF where, as part of the carbonic acid buffer system, its function is to regulate ECF pH. The carbonic acid/bicarbonate buffer system has traditionally been seen to be central to normal acid-base regulation, and details of the physiological role and importance of this buffer are given in Section 3.v. Suffice to say that bicarbonate can be envisaged as either a first- or second-line defence against an increase in proton concentration in the ECF. Being a base, bicarbonate can accept a proton to form weak carbonic acid, which then undergoes dissociation forming carbon dioxide and water:

$$HCO_3^- + H^+ \longrightarrow H_2CO_3 \longrightarrow CO_2 + H_2O$$

An alternative model (the 'second line') is to imagine that, because of their more favourable pK_a ranges, most proteins are negatively charged (polyanions) at blood pH and due to their relatively high

Essential Fluid, Electrolyte and pH Homeostasis, First Edition. Gillian Cockerill and Stephen Reed.

concentration, proteins are the initial buffers of unwanted protons. Such protons are then relayed to bicarbonate in order to regenerate the full negative charge on proteins;

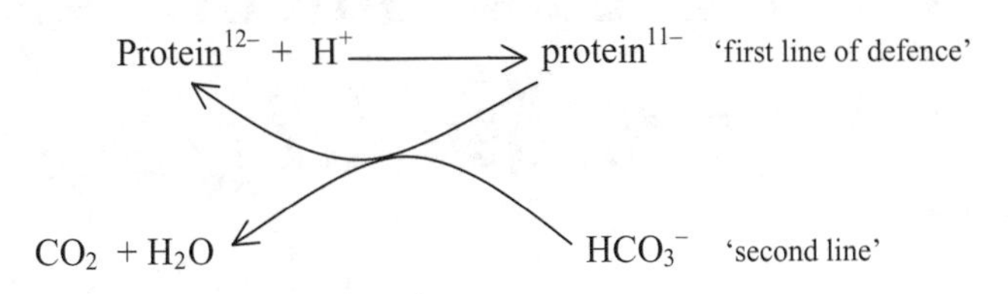

The net effect is the same regardless of the model: the 'free' proton has been incorporated into water at the expense of a bicarbonate ion.

Phosphate occurs in plasma in two ionic forms, HPO_4^{2-} and $H_2PO_4^-$, which together account for about 0.2 mmol/L, the remainder (0.6–0.9 mmol/L) being protein-bound. The relative proportion of the two ionic forms is fixed by the pH of plasma at about 4:1 (HPO_4^{2-} to $H_2PO_4^-$). Large amounts of phosphate are found combined with calcium in bone as hydroxyapatite. Inside cells (Table 1.3, Section 1.viii), phosphate is often combined with important metabolites such as adenosine triphosphate (ATP) and compounds within the glycolytic pathway (e.g. glucose-6-phosphate). Phosphate is the most important buffer in urine. The equivalent of about 70 mmol/L of acid is added to the glomerular filtrate each day as it passes along the nephron. Were this quantity of protons to be excreted in 1500 mL of unbuffered urine, its pH would be <1.5.

SECTION 2.xii

Self-assessment exercise 2.2

1. Which of the following statements is/are true? (there may be more than one correct answer)
   (a) ADH and aldosterone exert their effects throughout the length of the nephron;
   (b) ADH acts mainly in the proximal region of the nephron *and* aldosterone is effective mainly in the distal part of the nephron;
   (c) ADH acts mainly in the distal part of the nephron *and* aldosterone is effective mainly in the proximal region of the nephron;
   (d) ADH and aldosterone both operate only in the proximal region of the nephron;
   (e) ADH and aldosterone both operate only in the distal region of the nephron.
2. Aldosterone targets which of the following cell types?
   (a) α-intercalated cells
   (b) β-intercalated cells
   (c) principal cells
3. Name the two types of transport protein required to reabsorb sodium from the glomerular filtrate to the bloodstream.
4. Which of the following statements can be correctly used to complete the sentence '*Most* sodium reabsorption in the nephron...' (there may be more than one correct answer)
   (a) is independent of aldosterone;
   (b) occurs in the proximal tubules;
   (c) is carrier-mediated;

Essential Fluid, Electrolyte and pH Homeostasis, First Edition. Gillian Cockerill and Stephen Reed.

(d) is influenced by the glomerular filtration rate, i.e. the volume of fluid passing through the nephron.

5. A lack of aldosterone is likely to result in which of he following? (there may be more than one correct answer to this question)
   (a) an increase in the plasma concentrations of both Na^+ *and* K^+;
   (b) a decrease in plasma [Na^+] and an increase in plasma [K^+];
   (c) a decrease in the plasma concentrations of both Na^+ *and* K^+;
   (d) an increase in plasma [Na^+] *and* a decrease in urine[K^+];
   (e) an increase in urine [Na^+] *and* a decrease in urine [K^+].
6. Rare renin-secreting tumours are a cause of hypertension. TRUE or FALSE. Explain your answer.
7. Factors such as hypoxia (oxygen debt) or metabolic poisons that impair ATP generation will result in hyperkalaemia. Explain why this should be the case.
8. How can we explain mechanistically the link between a rise in blood pH and hypokalaemia?
9. Plasma concentrations of sodium, potassium, calcium, magnesium and iron all give a good indication of the total body content of each mineral. TRUE or FALSE. Explain your answer.
10. Which of the following is/are true of plasma iron? (there may be more than one correct answer to this question)
    (a) most Fe in the circulation is bound to ferroportin;
    (b) most Fe in the circulation is bound to transferrin;
    (c) most Fe in the circulation is bound to transferrin receptor;
    (d) most Fe in the circulation is bound to ferritin;
    (e) most Fe in the circulation is in the reduced form.
11. Comment on the following set of data obtained on a plasma sample.

Total calcium	2.18 mmol/L	(ref: 2.25–2.55 mmol/L)
Phosphate	1.0 mmol/L	(ref: 0.8–1.2 mmol/L)
PTH	5.0 pmol/L	(ref: 1–6 pmol/L)
Albumin	31 g/L	(ref: 35–50 g/L)

SECTION 2.xiii

Laboratory measurements 1: Osmometry

(i) Quantification of plasma or urine osmolality

We learnt in Section 1.iii that solutions and pure solvents exhibit different physical behaviours: the so-called colligative properties. A colligative property reflects the 'collective' or 'bulk' nature of a solution that arises due to the *number* of particles present in a solution rather than their exact chemical *nature*. Because urine, plasma and other body fluids are complex solutions of chemical species, the osmolality (a measure of the 'total solute content') of any of these fluids can be assessed by studying their colligative properties. Although proteins are present in high concentrations (typically g/L rather than mg/L), their large molecular mass means that there are relatively few individual molecules present compared with the numbers of ions, and thus it is these ions along with small molecular mass compounds that contribute most significantly to the osmolality. For example, a typical plasma concentration of albumin is 40 g/L and the molecular mass of this protein is approximately 65,000; the molar concentration of albumin in plasma is therefore approximately 0.6 mmol/L which compares with a normal plasma sodium concentration of 140 mmol/L.

Strictly speaking, the osmolality is a function of the thermodynamic *activities* of the chemical species present, indicated below as γ[ ] rather

Essential Fluid, Electrolyte and pH Homeostasis, First Edition. Gillian Cockerill and Stephen Reed.

than their molar concentration. Thus:

$$\text{Measured osmolality} = \gamma[\text{Na}^+] + \gamma[\text{K}^+] + \gamma[\text{HCO}_3^-] + \gamma[\text{Cl}^-] + \gamma[\text{glucose}] + \gamma[\text{urea}] + \cdots$$

Fortunately, from a quantitative point of view, for dilute solutions such as body fluids, $\gamma \sim 1$.

Routine laboratory measurements of osmolality are based on measurement of either the depression of freezing point or the elevation of vapour pressure of the sample under carefully controlled conditions.

Freezing point osmometry (cryoscopy) was first developed as a technique for use in the clinical laboratory in the late 1950s, and is best described by reference to cooling curves (see Figure 2.18) for both pure solvent and solutions. Assuming a steady rate of cooling, as the temperature of a solvent is reduced, a 'slush point' temperature is reached at which liquid and solid are in equilibrium; this temperature is maintained for a short while before the liquid forms a solid (Figure 2.18a). In the case of a solution however, the 'slush point' temperature, and thus the freezing point of the solution, is at a lower temperature (as shown in Figure 2.18b) than for the pure solvent.

In practice, a cryoscopic osmometer cools the sample solution to below its true freezing point and then initiates rapid freezing by a short period of vigorous stirring achieved by the rotation of a small probe. As ice crystals form in the sample, latent heat of fusion is liberated and the solution temperature rises to the 'slush point' just below the true

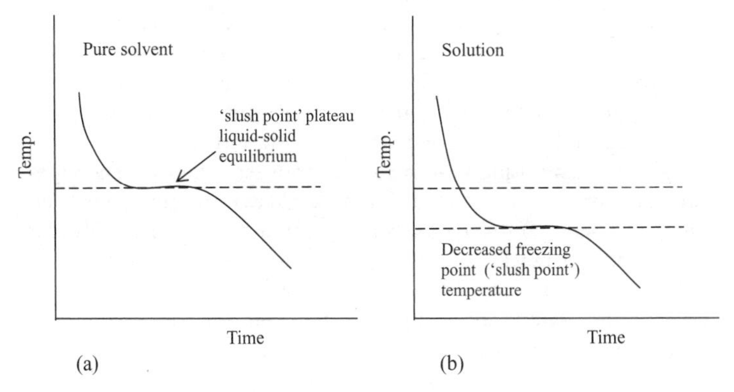

Figure 2.18 Cooling curves

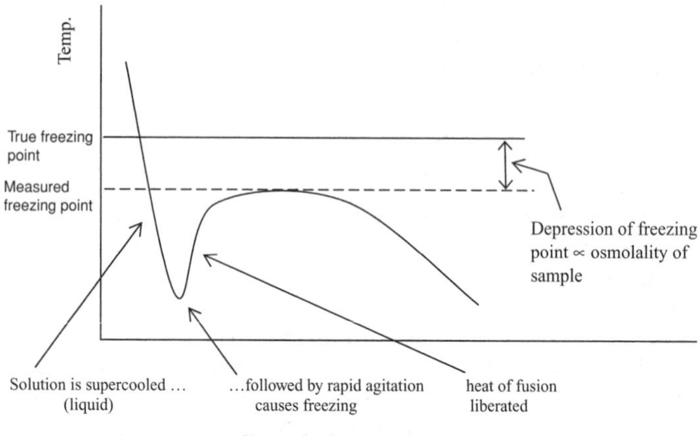

Figure 2.19 Cryoscopy

freezing point of the pure solvent. The numerical difference between the slush temperature and the true freezing point is a measure of the osmolality of the solution: the more solutes present in the sample, the greater the temperature difference (see Figure 2.19).

The commonly used alternative to cryoscopy for the estimation of osmolality is the measurement of changes in vapour pressure of the solution, although this is strictly more correctly measurement of the dew point temperature, i.e. the temperature at which water will condense from a wet vapour. The sample is introduced into a small chamber which is then sealed. Temperature and vapour pressure equilibria between the sample and the internal atmosphere will be reached within a few seconds before an electric current is passed through a thermocouple, bringing about a fall in temperature (Peltier effect) to below the dew point, and so water condenses onto the surface of the thermocouple probe, releasing heat of condensation as it does so. The slight rise in temperature is detected and expressed as osmolality. One main difference between this type of instrument and the cryoscope is that dew point devices will not detect volatile substances such as alcohol.

For practical reasons, in dilute solutions such as body fluids, the effective osmolality can be approximated from the total solute concentration; the osmolarity of a solution is an estimate of the total concentration of all of the compounds present. However, recall from Section 1.iii that although urea is measured *in vitro* as a component of the total solute content of plasma, it has little real osmotic effect on most cell membranes *in vivo* because it is a freely diffusible molecule. Uraemia as a consequence of, say, renal dysfunction, internal bleeding or other causes of increased protein turnover, will give spuriously high measured osmolality results, but the true tonicity of the sample due to the concentrations of non-diffusible ions and molecules is in fact normal.

Remember, it is the concentrations of sodium and protein (albumin) that are the principal determinants of fluid flux between ISF and ECF, and plasma and ISF compartments respectively. Abnormalities in sodium concentration will affect the state of cellular hydration and volume, whilst low plasma albumin concentrations will result in accumulation of fluid within tissue spaces, causing an increase in ISF volume (oedema). Thus, accurate measurements of plasma [Na^+] and osmolality are important to assess fully a patient's state of cellular hydration, whereas accurate estimation of albumin concentration is required in cases of oedema.

Samples for osmolality measurement can be safely stored for up to a week at 4°C, but must be kept tightly capped as evaporation of plasma water will increase the concentration of the sample.

SECTION 2.xiv

Laboratory measurements 2: Ion selective electrodes (ISEs)

Traditionally, quantification of ion concentration in biological samples was performed using a technique known as flame photometry in which the intensity of light emitted as the sample was burnt in a flame was measured. The 1980s, however, saw the progressive introduction into the clinical laboratory of ion selective electrode technology. Ion selective electrodes (ISEs) are electrometric devices, measuring electrical signals (voltage or current) generated in response to the type and number of ions present in the sample. Similarly designed devices, more correctly called 'gas probes', are used to measure the pressures of oxygen, using the Clark electrode, and carbon dioxide using the Severinghaus electrode. The latter is a modified pH ISE probe.

A major advantage of ISEs over older flame-based methods is that electrodes can be conveniently used at the bedside for 'point-of-care testing' (POCT). A fully automated 'multi-test' analyser will rapidly generate data for Na^+, K^+, pH, PCO_2 and PO_2 simultaneously. Table 2.4 below shows commonly used examples of electrometric analyses.

Ion selective electrodes

Somewhat confusingly perhaps, an ion selective electrode (ISE) actually consists of *two* metal wire electrodes, usually made of platinum, one of which is in contact with the sample and the other contacts

Essential Fluid, Electrolyte and pH Homeostasis, First Edition. Gillian Cockerill and Stephen Reed.

Table 2.4 Electrometric methods

Technique	Electrical signal measured	Clinical applications
Potentiometry	Voltage at constant current	H^+, Na^+, K^+, Ca^{2+}, NH_4^+, Cl^-, carbon dioxide
Amperometry	Current at constant voltage	Oxygen

an electrolyte solution of fixed ionic composition which acts as a reference. The sample and the electrolyte solution are separated by an analytical membrane (not a biological membrane), which may be made of glass that has been specially formulated to be selectively responsive to only one type of ion, for example, the pH membrane is selective in its response to H^+. Analytical membranes may also be constructed of an ionophore such as valinomycin or crown ether embedded in a neutral PVC matrix.

An electrical signal is measurable when an electrochemical 'cell' is set up; a cell consists of:

- An indicator (or sensing) electrode with a selectively responsive membrane made of glass or a synthetic polymer;
- A reference electrode, usually glass;
- A meter to measure the signal generated; and
- A solution, i.e. a sample, containing the ion of interest.

The basic construction of a typical ISE, which is about the size of a small reel of cotton thread, is illustrated in Figure 2.20a and b.

Typical chemical structures of ionophores are shown in Figure 2.21. First discovered in 1967, crown ethers are cyclic compounds which are able to engage with ions in a 'lock and key' like fashion; certain ions neatly fit into the lumen of the crown. Early potassium ISEs used as their ionophore valinomycin, a natural product that binds fairly specifically with K ions, whilst the original Na ISEs used Na-responsive glass membranes; both types have now been superseded by appropriately selective crown ethers.

The two electrodes that make up an ISE 'cell' are sometimes referred to as half-cells. Critically, the indicating electrode consists of an 'electroactive membrane' which is responsible for detecting the presence and concentration of ions in the sample. In order to complete the circuit, there must be an electrical connection between the reference electrode and the sample solution: this is called the liquid

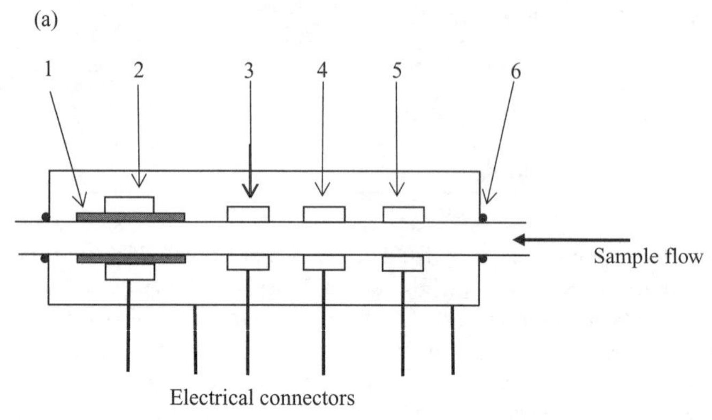

Key:

1. ceramic liquid junction
2. reference electrode
3. chloride electrode
4. potassium electrode
5. sodium electrode
6. O-ring seal

(b)

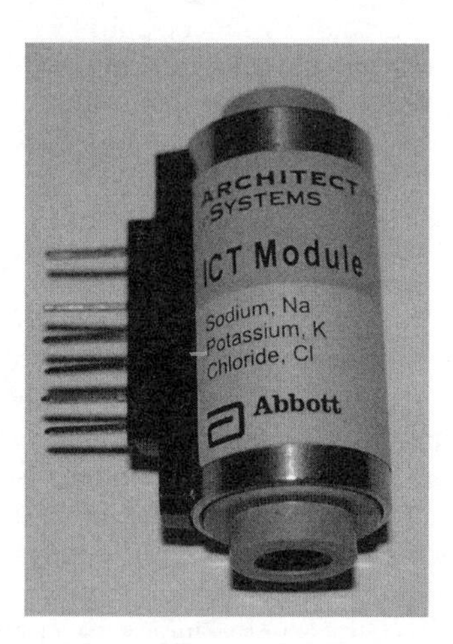

Figure 2.20 (a) Schematic of an ion selective electrode assembly; (b) Ion selective electrode module. (Adapted and reproduced with permission of Abbott Inc.)

Ionophore

Alkyl O O O O Alkyl

P P

Alkyl O O — Ca — O O Alkyl

Alkyl Phosphate

Mediator

$C_8H_{17}O$ O

P

C_8H_{17} C_6H_5

di-*n*-octylphenyl phosphonate

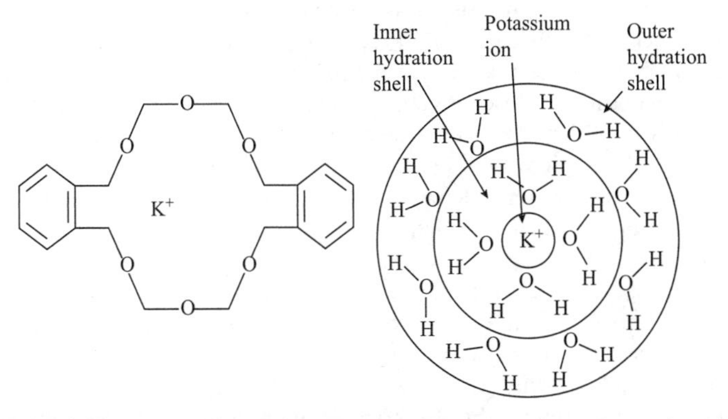

Figure 2.21 Ionophores: valinomycin and crown ethers. Reproduced with permission from Burnett, D. and Crocker, J. (2005) *The Science of Laboratory Diagnosis* (Wiley-Blackwell).

junction. An electromotive force (EMF) is developed at the boundary of two dissimilar electrolyte solutions, such as at a liquid junction, thus a liquid junction potential (LJP) is defined as the voltage that arises due to charge separation at the interface of two solutions of different ionic strength and/or charge due to differences in the relative rates of diffusion (mobilities) of the ions present.

The actual potential developed at an electrode (half-cell) cannot be measured in absolute terms and can only be expressed relative to another half-cell, hence the need for a reference electrode. Mathematically, where E represents the potential;

$$E_{cell} = E_{ind} - (E_{ref} + E_J) \quad (1)$$

where E is the potential; ind, indicating electrode; ref, reference electrode, J, liquid junction.

Because the reference electrode is not responsive to any particular ion, the potential developed, E_{ref}, is independent of ion activity so it is in effect a numerical constant for any given ISE device. Furthermore, the LJP can be assumed to be constant, so Equation 1 simplifies to:

$$E_{cell} \propto E_{ind} \quad (2)$$

Potentiometry is the most commonly used electrometric technique in routine clinical analysis for assessment of fluid and electrolyte disturbances. Measurements are precise (under controlled conditions), often showing less than a 1% variation; ion estimation is possible in coloured, turbid or viscous samples and even in whole blood. The high sensitivity of the technique means that measurements are possible in very small sample volumes (typically 25 μL of plasma) and electrodes are responsive over a wide range of concentrations (e.g. 10^{-2} to 10^{-6} molar), greater than the physiological range for commonly measured ions.

Because of differences in the chemical compositions of the reference electrolyte solution and the sample, a small voltage (measured in millivolts) develops across the glass membrane. The actual value of the voltage is determined by the ionic composition of the sample, that is, the voltage is related to the concentration of that particular ion.

In reality, ISEs measure not ion concentration but ion activity, which relates to the probability of two ions interacting with each other in solution. Ion activity is determined by the nature of the matrix that contains the ions, for example, total ionic strength and temperature both affect the ion activity. Because of the complex chemical nature of plasma, ion activity probably gives a more physiologically useful estimate of electrolyte status than does ion concentration. The relationship between ion activity and concentration is given by:

$$C_x = \frac{a_x}{\gamma} \quad (3)$$

where C_x is the concentration of ion 'x', a_x is its activity, and γ is the ion activity coefficient.

The activity coefficient, γ, has a numerical value that varies somewhat between different ions, but approaches 1.0 in dilute solutions, so for practical purposes, $C_x \approx a_x$. Although ISEs respond to ion activity rather than concentration, in most clinical situations the two can be considered equivalent. However, it should be noted that the actual value of γ is not a constant and varies with ionic strength and ionic composition of the solution being tested and the nature of the sample matrix, dependent mainly upon the protein and lipoprotein content of the plasma, and pre-dilution of each sample in an ionic strength adjustment buffer (ISAB) may be required (see also the discussion on direct and indirect measuring devices below).

The potential developed at the indicating electrode is, as stated above, determined by the activity of the ion of interest and is defined by the Nernst equation:

$$E = E^0 + \frac{RT}{nF} \ln a_x \qquad (4)$$

where E is the measured potential in volts; E^0 is a constant that is characteristic of the particular ISE (in volts); R is the gas constant (8.314 joules/degree/mole); T is the absolute temperature; n is the valency (charge) of the ion being quantified; F is the Faraday constant, 96,500 coulombs; and $\ln a_x$ is the natural logarithm of the activity of ion 'x'.

This may be re-written with $\log_{10}$ replacing natural logs:

$$E = E^0 + 2.303\frac{RT}{nF} \log_{10} a_x \qquad (5)$$

At 25°C, for a monovalent ion such as Na^+ or K^+ where $n = 1$ and substituting values for the constants, this version of the Nernst equation simplifies to:

$$E = E^0 + 0.0591 \times \log_{10} a_x \qquad (6)$$

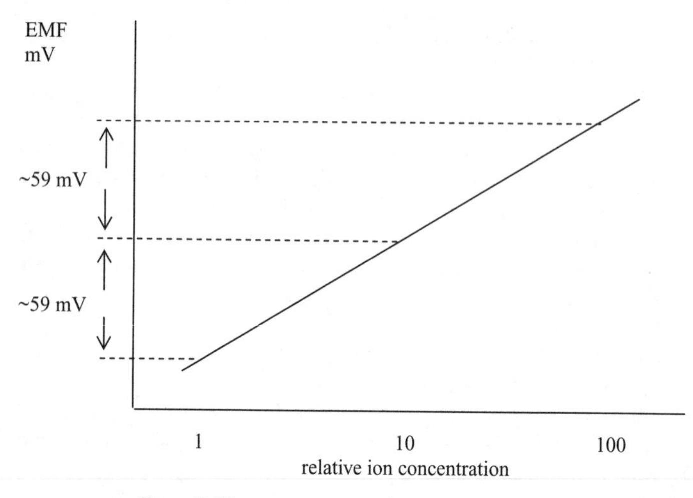

Figure 2.22 Logarithmic response of an ISE

Rearranging Equation 6 gives

$$E = 0.0591 \times \log_{10} a_x + E^0 \tag{7}$$

which is clearly an equation for a straight line, $y = mx + c$ with a slope of 0.0591 volts (Figure 2.22). Expressed another way, each order of magnitude (10-fold) change in ion activity corresponds to a theoretical voltage change of 0.0591 volts (or 59.1 mV).

Throughout the discussion so far, the term ion *selective* rather than ion *specific* has been used. Analytical membranes are not chemically specific and will respond not only to the primary, or determinand, ion of interest but also to an interfering ion. The relative responsiveness of the membrane is defined by its selectivity coefficient, $k_{x,y}$ where x and y stand for unspecified primary and interfering ions respectively;

$$k_{x,y} = \frac{\text{Response to y}}{\text{Response to x}} \tag{8}$$

The Eisenmann-Nikolsky equation is an expanded version of the Nernst equation that takes into account the selectivity coefficient of the electrode:

$$E = E^0 + 0.0591 \times \log_{10}\{a_x + (k_{x,y}.a_y)\} \tag{9}$$

where a_x is the activity of the primary ion x and a_y is the activity of the interferant ion y.

For electrodes that have very high selectivity (i.e. low cross-responsiveness to interfering ions), $k_{x,y}$ is a very small number and so too therefore will be the product ($k_{x,y} \cdot a_y$). Quoted values for $k_{x,y}$ for commercial ISEs are:

for	Na,K	2×10^{-2}
for	K,Na	2.5×10^{-3}
for	Cl,Br	2.0

Direct or indirect analysis?

Ion selective electrode devices are described as 'direct' (when used to measure ion concentration in samples of *un*diluted plasma or even whole blood) or 'indirect' (the blood cells must be removed and only the plasma is analysed after it has been diluted, typically 1:20 or 1:30). Both types of instrument are robust and reliable, capable of producing results with high accuracy and precision, but indirect ISEs are subject to error if samples contain very high concentrations of proteins

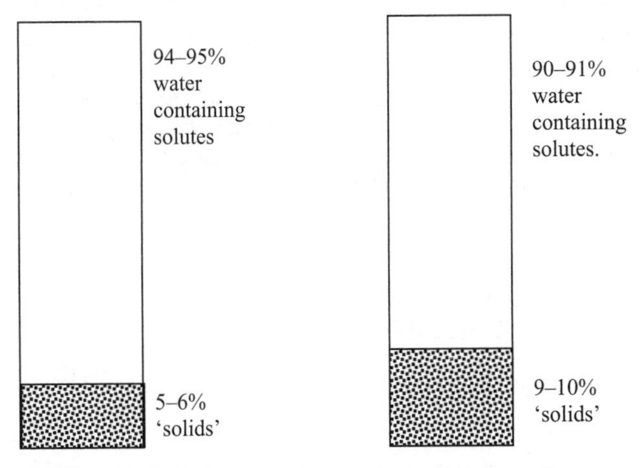

Figure 2.23 Fluid displacement by proteins and lipoproteins
The $[Na^+]$ in the *water phase* is the same in both cases but an increase in the proportion of colloids causes a sampling error in indirect analysis.

or lipids. The error arises because indirect measurements require the plasma samples to be pre-diluted in a buffer solution before analysis. Large colloidal molecules and lipoprotein complexes take up space and so effectively displace plasma water, which is of course where the electrolytes are located (Figure 2.23). Thus, for a typical sample, 100 μL of plasma equates to approximately 95 μL of water, but in the case of a hyperlipaemic or hyperproteineamic sample, 100 μL may equate to only 90 μL of plasma water, so the sample will be overdiluted when mixed with the buffer. For example, suppose 100 μL of plasma with normal protein and lipid concentrations is mixed with 2.9 mL of buffer. The apparent dilution is 1/30, but if we consider only the plasma water component, the dilution is 95 μL in a total of 3 mL, equivalent to a 1/31.6 dilution. For a sample with elevated proteins or lipid concentrations, the real dilution is 90 μL in 3 ml which is a 1/33.3 dilution. The consequence of this overdilution of the plasma water results in erroneously low values (by approximately 5%); most noticeable is usually a low sodium concentration. Because this is an analytical artefact and not a true physiological state, the finding is described as pseudohyponatraemia. In other words, the sodium concentration in plasma water is actually normal, but the sampling error means that a smaller volume than expected of plasma water has been analysed, so therefore a reduced concentration of Na^+ has been detected.

On the other hand, indirect ISEs have advantages over direct measuring devices: (a) they require a smaller sample volume than do direct instruments; and (b) pre-dilution in a suitable buffer solution will standardise the analysis by controlling pH and, more importantly, the ionic strength of the solution.

SECTION 2.xv

Laboratory measurements 3: Calcium, magnesium, vitamin D, phosphate and iron

Concentrations of total calcium in plasma or urine can be quantified by atomic absorption spectrophotometry (AAS), or more conveniently by an automated colorimetric assay based on complex formation with a dye. Atomic absorption is the reference method but is labour-intensive relative to colorimetry. Samples for AAS analysis are diluted in a strontium-containing solution to prevent possible interference by phosphate.

Commonly-used calcium chelating dyes are cresolphthalein complexone (CPC), arsenazo III or chlorophosphonazo III (CPZ). In all cases, the intensity of the colour formed due to the dye–calcium interaction is proportional to total calcium concentration. Possible interference by magnesium in the CPC method may be overcome by the addition of a chelating agent such as 8-hydroxyquinoline to the colour reagent. Samples of serum or plasma are equally suitable, but blood collected into EDTA (ethylenediamine tetra acetic acid) may not be used as it binds calcium, giving falsely low results. Plasma ionised calcium concentration, the physiologically active fraction of the total, is measured using ISE technology. Correct sample handling is important when measuring [Ca^{2+}]. Blood must not be collected into a chelating anticoagulant such as oxalate, citrate or EDTA; the assays should

Essential Fluid, Electrolyte and pH Homeostasis, First Edition. Gillian Cockerill and Stephen Reed.

be carried out as quickly as possible, and if possible maintained in anaerobic conditions since changes to calcium-albumin binding may occur as loss of CO_2 from the sample to the atmosphere will cause the pH to rise. Urine for calcium estimation should be collected for a full 24-hour period into a bottle containing 6 M hydrochloric acid to prevent precipitation of calcium phosphate on storage.

Plasma magnesium is estimated using atomic absorption or dye-binding methods. A commonly used reaction is that between the diazonium salt xylidyl blue and magnesium at pH 11. Unusually for colorimetric assays, the reaction between them causes a reduction in absorbance compared with that of the dye alone. Any possible cross-reaction between the diazonium dye and calcium is prevented by the addition of a chelating agent ethylene glycol tetra acetic acid (EGTA). The method may be used for plasma (blood collected into lithium heparin anticoagulant) or serum. Whole blood that is not separated into plasma and cells quickly, or haemolysed samples, are not suitable for analysis due to the relatively high magnesium concentration in intracellular fluid, which leaks into ECF when red cells are damaged.

Phosphate is routinely measured in plasma, serum or urine using molybdenum-containing reagents. The reaction is not well defined but involves reduction, producing phospho-molybdenum complex which has an intense blue colour and is thus easily quantified colorimetrically.

Assessment of vitamin D status is problematic because of the different chemical forms of the vitamin. The main circulating metabolite is 25-hydroxy vitamin D3 (25-OH D3), although treatment of deficiency is usually with 25-hydroxy vitamin D2 (ergocalciferol) and this is widely used as a food supplement. Many laboratory methods measure total vitamin D, designated 25(OH)D, but wide variability between immunoassay methods, a lack of adequate standardisation, and uncertainty as to their ability to recognise both vitamin D2 and D3 have cast doubts on the reliability of the assays for the purposes of both diagnosis of deficiency and monitoring of treatment.

As would be expected of a steroid-like compound, 25-OH D3 is bound to a plasma globulin (vitamin D binding protein, VDBP) for carriage through the bloodstream and this binding must be broken before analysis can continue. Commonly used methods are based on immunoassay, high-performance liquid chromatography (HPLC) with either ultra-violet detection or HPLC with detection and quantification by mass spectrometry (LC-MS). Definition of a reference range for vitamin D has proven to be difficult. Plasma values of less than 25 nmol/L are associated with clinical signs of vitamin D deficiency, and values greater than 75 nmol/L are considered to indicate sufficiency. Values falling into the intervening range 25–75 nmol/L are difficult to interpret as this represents a 'grey area' of diagnostic uncertainty.

Measurement of serum iron is also somewhat problematic. Methods involve chelation of the iron with complex cyclic molecules such as triazines which form intense colours. However, the risk of contamination of the sample or of the reaction vessels is high, serum iron concentration is a poor indicator of total body iron status because most of the mineral is intracellular, plasma values vary during the course of the day, and a number of chronic inflammatory conditions will decrease serum iron concentration independently of any pathological change in the iron status. Most clinical laboratories offer estimation of serum transferrin, soluble transferrin receptors and ferritin, all measured by immunoassay as markers for iron overload or deficiency.

SECTION 2.xvi

Laboratory measurements 4: Miscellaneous methods for clinically useful analytes

(i) Flame-based methods for cations

Prior to the introduction of ISEs, routine analysis of plasma and urine for electrolyte concentration was based on the use of flame emission photometry. A simple means to identify a cation in a sample is to place a small amount into a flame and observe the colour change that occurs; sodium burns with a yellow flame, potassium with a purple/violet flame and calcium with a red flame. The intensity of the emission so produced is related to the concentration of the ion present, and a flame photometer is simply a device that measures the intensity. This technique, which was really only suitable for quantification of monovalent cations such as Na^+, K^+ and Li^+, is now essentially redundant in modern clinical laboratories, but flame absorption (or atomic absorption) spectrophotometry is very much in use for calcium, magnesium and many trace elements.

In atomic absorption, an atomic vapour is formed by spraying a dilute solution of a plasma sample into a flame. The flame plays no direct

Essential Fluid, Electrolyte and pH Homeostasis, First Edition. Gillian Cockerill and Stephen Reed.

part in the analysis, but merely acts to 'contain' the vapour within a defined space. Quantification is achieved when electromagnetic radiation (EMR) produced by an ion-specific cathode lamp is passed through the vapour. The energy of the EMR is matched exactly to that required to cause electrons of the atoms being measured to undergo excitation to a higher orbital. In effect, some of the energy of the EMR has been 'given' to or absorbed by the atoms in the vapour and there is a reduction in the intensity of EMR that emerges from the flame. The amount of energy absorbed by the ions during the excitation process is proportional to the concentration of the analyte atom.

Atomic absorption methods are not easily automated, so for a busy diagnostic laboratory such assays are not very convenient when large numbers of samples need to be analysed each day.

(ii) Albumin and total protein estimation

Plasma albumin is easily quantified by a simple colorimetric assay. The protein combines with a dye such as bromocresol green (BCG) or bromocresol purple (BCP). Both of these dyes are pH indicators that change colour in the presence of protein. The magnitude of the colour change, as shown by the intensity of the colour, is proportional to albumin concentration. These methods are inexpensive and conveniently automated, but the BCG method is subject to interference ('cross-reaction') by plasma globulins. When albumin concentration is low, the relative interference is significant, so, ironically, in just those clinical situations when accurate measurement of albumin is required, the method is unreliable. The BCP method is less prone to globulin interference. A far more reliable but more expensive method is to measure albumin using an immunoassay. A specific anti-human albumin antibody combines with albumin (the antigen in this case) in the patient's sample to form an antigen-antibody (Ag-Ab) complex. The extent of binding gives an accurate estimate of albumin concentration.

Estimation of plasma total protein (TP) is of limited value in most clinical practice, but it may have a role to play in assessing a patient's state of hydration. Because protein molecules are large and tend not to leave the blood vascular system, any reduction in plasma volume will cause an increase in [TP] and conversely an expansion of plasma volume will cause [TP] to decrease. The most commonly used method for quantifying [TP] is the biuret reaction in which Cu^{2+} ions in alkaline solution form a complex with peptide bonds within the proteins. This results in a colour change from bright blue ($CuSO_4$) to purple (the complex between Cu and protein is in fact red, but this is masked by the blue colour of the biuret reagent).

(iii) Urea

Knowledge of the plasma concentration of urea is an indication (albeit a fairly poor one) of dehydration or of renal insufficiency, both of which may be coincident with fluid and electrolyte imbalance. Interpretation of osmolality results may be helped if the urea concentration is known, because urea being able to diffuse freely across cell membranes, the solute contributes to the osmolality (total concentration of solutes) but does not affect the *in vivo* tonicity. Most laboratories nowadays measure urea in both plasma and urine using an enzyme-based method. Urease catalyses the following reaction:

$$H_2N\text{-}\underset{\underset{O}{\|}}{C}\text{-}NH_2 + H_2O \longrightarrow 2NH_3 + CO_2$$

Measurement is made by quantification of the NH_3 (or rarely CO_2) produced from this reaction. Ammonia (or if the solution is acidified, NH_4^+) can be measured by ISE-type technology.

(iv) Glucose

Fluid and electrolyte disturbances are quite common in diabetics and, like urea, knowledge of plasma glucose concentration may help with the interpretation of osmolarity results. Glucose is measured in plasma using methods that use either glucose oxidase/peroxidase (GO/POD) or glucokinase/glucose-6-phosphate dehydrogenase (GK/G6PD).

$$\text{Glucose} + O_2 \longrightarrow \text{gluconic acid} + H_2O_2 \quad \text{step 1}$$

$$H_2O_2 \longrightarrow H_2O + O_2 \quad \text{step 2}$$

Oxygen consumption at step 1 or oxygen liberation at step 2 is used as the measurement end point.

Alternatively, a coupled reaction with hexokinase (HK) and glucose-6-phosphate dehydrogenase (G6PD) can be used as shown below

$$\text{Glucose} + \text{ATP} \longrightarrow \text{glucose-6-phosphate}$$

$$\text{Glucose-6-phosphate} \longrightarrow \text{6-phosphgluconate}$$

$$NADP^{+} \rightarrow NADPH + H^{+}$$

The rate of reduction of $NADP^+$ is a measure of the initial glucose concentration. The increase in NADPH concentration is monitored by spectrophotometry at a wavelength of 360 nm.

SECTION 2.xvii

Self-assessment exercise 2.3

1. You are able to select from two possible ISEs for the estimation of Na in plasma. For one the Eisenmann-Nickolsky is quoted as 5.2×10^{-5} and for the other the same parameter has a value of 8.7×10^{-7}. Which of the two would you choose to use; give reasons to support your answer.
2. The reference range for pH of arterial whole blood is 7.35–7.45. What would be the difference in electromotive force (EMF) generated by the ISE at these two limits?
3. The measured osmolality of a plasma sample is 330 mmol/kg, yet the calculated osmolarity is 305 mmol/L. How can the difference be explained? (*Hint: review Section 1.xiii*)

Check your answers before proceding

Essential Fluid, Electrolyte and pH Homeostasis, First Edition. Gillian Cockerill and Stephen Reed.

Disorders of fluid and electrolyte balance

SECTION 2.xviii

Introduction

Disorders of fluid and electrolyte balance are very commonly encountered in routine clinical practice, and their treatment often represents a clinically urgent situation. Reliable laboratory data are therefore essential for correct treatment, but even so, it is estimated that a significant number of cases of inappropriate treatment, some with serious consequences, occur each year. In general terms, imbalances in water and electrolytes arise due to:

- Reduced intake or reduced absorption in the gut
- Increased losses via gut or kidney
- Increased intake (rarely)
- Redistribution between body fluid compartments
- Drugs and other iatrogenic causes (as a result of therapy)
- Genetic factors.

Reduced intake of water or minerals is not likely in someone following a typical diet, but personal dietary preferences or food sensitivities, pre-existing illness or on occasions the unavailability of food or water may all dispose towards poor intake. The absorption of certain minerals may be affected by other constituents within the diet. For example, a diet rich in cereals containing organic phosphates called phytates or vitamin D deficiency may cause reduced absorption of calcium from the gut, whereas alcohol consumption can modify iron absorption.

As described in the foregoing sections, hormones play a major role in regulating electrolyte loss mainly via the kidney. Thus endocrine or renal disorders may lead to changes in the excretion of water or electrolytes. Renal losses may be the intended result of drug use (e.g. diuretics), but poor monitoring can lead to losses beyond those desired or due to nephrotoxicity. Diarrhoea and/or vomiting can quickly lead to dehydration and the loss of electrolytes. In contrast, unless there is

Essential Fluid, Electrolyte and pH Homeostasis, First Edition. Gillian Cockerill and Stephen Reed.

some underlying pathology such as renal disease, excessive intake of electrolytes, minerals or water is unlikely to cause overload.

Maintenance of normal blood volume and pressure are vital. Section 2.iii outlines key homeostatic responses to hypovolaemia, but where homeostasis is disrupted through illness or injury, intravenous rehydration fluids are necessary to expand the volumes of body compartments.

Intravenous infusions are of two broad types: (i) isotonic crystalloid salt solutions and (ii) colloids. Examples of crystalloid salt solutions are NaCl ('normal' 0.9% saline), dextrose-saline and Hartmann's solution. Sometimes additional solutes need to be added to treat particular conditions, for example, sodium bicarbonate infusion is used in cases of acidosis. Colloids are usually composed of a protein such as gelatine or albumin or a polysaccharide such as hydroxy-ethyl starch.

Fluids become distributed into body fluid compartments according to their diffusibility and the relative size of each compartment. Because of their high molecular weight, colloids are restricted to the intravascular (plasma) compartment; dextrose-containing solutions are distributed throughout the total body water space so have relatively

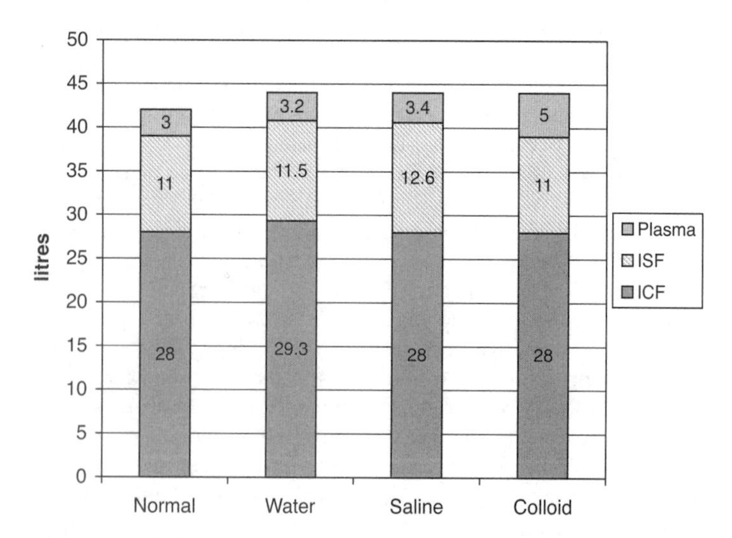

Figure 2.24 Changes in compartment volume following infusion of 2 litres of fluid

little effect on plasma volume. Sodium-containing crystalloids are distributed to the ECF space, but because the plasma volume is a small proportion of total ECF, the effect as a plasma expander is relatively slight. Figure 2.24 illustrates the distribution of 2 litres of infused fluid.

Inappropriate intravenous (i.v.) infusions can have undesirable effects. One survey estimated that in one year alone, over 8000 deaths in the U.S. were associated with the injudicious use of i.v. therapy, so regular monitoring of plasma values and clinical assessment are essential.

Arieff (1999) Chest **115**:1371–1377

SECTION 2.xix

Principles of data interpretation

In order for a correct interpretation of any laboratory result to be made we must ensure that the method of analysis is as reliable as it possibly can be, and that the sample is in 'as life-like condition' as possible when it arrives in the laboratory. All laboratory results are potentially subject to error due to specific problems occurring during the pre-analytical, analytical or post-analytical phases. Technical and clinical scientists in the laboratory carry out a range of quality control checks to ensure that, as far as is practicable, results being sent to clinicians on the ward or in the clinic are reliable and clinically valid, but a very few errors go unnoticed by laboratory staff. Furthermore, once the requesting doctor has the result of a test, s/he must then decide if it lies within normal limits of biovariation (the population-based 'reference range' for the analyte), or whether pathology is indicated. This Section deals with aspects of data interpretation and highlights some potential pitfalls.

Factitious results

Valid interpretation of results requires that the samples are a true reflection of the patient's physiological and biochemical status. A well-established dictum of laboratory diagnosis states that 'no result is better than the sample from which it was derived'. A badly-drawn blood sample, an incomplete collection of urine, or delay in sending samples to the laboratory will all have detrimental effects on the reliability of data produced. In addition to such *in vitro* changes in sample quality,

Essential Fluid, Electrolyte and pH Homeostasis, First Edition. Gillian Cockerill and Stephen Reed.

Table 2.5 Causes of factitious results

Laboratory result	Possible cause of invalid result
High plasma potassium concentration	(a) Haemolysed sample (b) Delay in separating cells from plasma (c) Contamination of sample with potassium salt of EDTA (ethylenediamine tetra acetic acid) anticoagulant (d) Samples collected in GP surgeries during summer months
High plasma sodium concentration, all other values are low	Patient with saline drip. Blood may have been taken from a drip-arm where the local Na (and Cl) concentration is high but all other constituents are diluted
Low plasma sodium concentration	Pseudohyponatraemia due to hyperproteinaemia or hyperlipidaemia
Very low plasma calcium concentration, often with a very high potassium concentration	Blood may have been collected into an anticoagulant such as EDTA (ethylenediamine tetra acetic acid)
Low plasma calcium concentration	Hypoalbuminaemia
Low plasma bicarbonate concentration, especially if measured as 'total CO_2' (TCO_2)	Plasma left exposed to the atmosphere for long periods. Because the PCO_2 of blood is much higher than it is in the atmosphere, CO_2 gas diffuses out of the plasma so reducing the bicarbonate concentration, by 'pulling' the equilibrium to the right the equation shown here: $HCO_3^- + H^+ \rightleftharpoons H_2CO_3 \rightleftharpoons H_2O + CO_2$

errors in interpretation may arise from *in vivo* effects that may have an impact on interpretation of a final result (e.g. the presence of any unsuspected pathophysiological changes which happen to be coincident with any apparent abnormality). Table 2.5 lists some common problems.

Pseudohyperkalaemia

Spuriously high plasma potassium results are believed to be fairly common but are not always easy to identify. There are a number of pathophysiological causes such as leucocytosis (increased white cell count) and increased platelet count, but a misleading K^+ result is more often due to pre-analytical factors. The most obvious cause is haemolysed samples; quite simply, damage to red cells causes potassium to leak into the plasma. Even quite small amounts of haemolysis can cause the plasma to appear pink/red due to the presence of free haemoglobin, yet the potassium value may be only mildly elevated, e.g. around 6 mmol/L. Plasma sodium concentration may be a little low in haemolysed samples due to dilution by intracellular fluid. Haemolysis becomes a problem only when using direct-measuring ion selective electrodes (see Section 2.xiv) because whole blood need not be centrifuged to separate cells from plasma, so any haemolysis will not be seen by the analysts.

Delay of even three to four hours in separating cells from plasma is a common cause of factitious potassium results, and often results in measured potassium values of around 10 mmol/L. As glucose concentration in the stored sample falls, due to its metabolism by cells the Na/K ATP'ase pump fails and K^+ leaks out of blood cells. Despite the very high values encountered, plasma may not appear pink/red, i.e. there need not be overt haemolysis. Biomedical scientists often notice that blood samples collected in GP surgeries or other clinics away from the main laboratory have elevated potassium results. This phenomenon is particularly apparent during the summer months when ambient temperatures are higher than during the winter, permitting more rapid metabolism by cells within the sample. The probable cause is, as above, most likely due to exhaustion of glucose in the sample leading to failure of the sodium ATP'ase pump. As potassium is clearly not the only constituent of intracellular fluid, spurious elevations are also seen with, for example, plasma magnesium concentration and the plasma activity of the enzyme lactate dehydrogenase.

Correct venepuncture technique is essential to ensure valid potassium results. It is common practice for phlebotomists to ask patients to flex (open and close) the fist or to grip a soft 'squeeze ball' whilst the blood is being drawn. Both practices have been shown to cause pseudohyperkalaemia.[2] In many instances, old-fashioned hypodermic syringes have been replaced with vacuum tubes for blood collection. A needle is inserted into a vein and an evacuated tube pushed on to the needle; one needle can be used to collect several tubes of blood.

---

[2] Bailey and Thurlow (2008) *Ann. Clin. Biochem.* **45**: 266–269.

The possibility of an error arises if the first tube to be attached to the needle contains K_2EDTA anticoagulant (for haematological tests). It is assumed that some degree of regurgitation of blood, now contaminated with EDTA, occurs into the needle as the blood flows into the evacuated tube, so when the second tube (for chemical analysis) is attached to the needle the traces of K_2EDTA are washed into the sample. The consequence is the 'chemistry' sample has an inappropriately high potassium concentration but low results are found for calcium, magnesium and zinc and alkaline phosphatase (which needs a divalent ion as activator). In one published study,[3] the authors found that nearly a quarter of samples with a potassium result of greater than 6.0 mmol/L were contaminated with K_2EDTA. Collecting samples in the correct 'draw order' is therefore imperative if contamination is to be avoided.

Pseudohyponatraemia

Plasma samples containing an abnormally high concentration of proteins or lipoproteins when analysed by indirect methods (i.e. those which require a pre-dilution of the sample) often appear to have a low $[Na^+]$. In truth, the sodium concentration in plasma water is usually normal, but errors arising during pre-dilution mean that a smaller volume than expected of plasma water has been analysed, so an incorrectly reduced concentration of Na^+ is reported. Although the effect is most noticeable with $[Na^+]$, clearly the concentrations of all analytes within the water phase of the sample will be similarly affected whenever there is a pre-dilution step in an assay method (see page 145).

Hypocalcaemia

The results for plasma *total* calcium concentration are low when the patient has hypoalbuminaemia. However, the physiologically active ionised calcium (Ca^{2+}) fraction may be normal. Check by calculating the 'corrected calcium' thus:

$$\text{corrected}\,[\text{Ca}] = 0.02 \times (41 - [\text{albumin}]) + \text{measured total}\,[\text{Ca}]$$

(*Note*: The value of 41 represents a normal plasma albumin concentration in g/L, but sometimes this equation is written with 40 replacing 41.)

---

[3] Cornes *et al.* (2008) *Ann. Clin. Biochem.* **45**: 601–603.

Urine analysis

Urine samples are no less prone to sampling, transport or storage errors. In many situations, especially in balance studies when output or loss of, say, an electrolyte is being assessed against intake, a complete 24-hour collection of urine is needed. Experience has shown that patients and even ward staff have difficulty in collecting the sample accurately. Furthermore, there has been debate around the need to acidify 24-hour urine collections for, in particular, calcium analysis. Traditionally, a few millilitres of hydrochloric acid would have been placed into a 2 litre plastic collection bottle prior to its despatch to the ward or out-patient clinic. Clearly, this has health and safety implications for the nursing staff and the patient. Some laboratories prefer to add acid upon arrival of the urine collection in the laboratory; others use no preservative at all.

Monitoring plasma values

Clinical laboratory results serve two main purposes: to confirm or refute an initial, possibly tentative, diagnosis, and secondly to monitor patients' response to treatment. Correct diagnosis is reliant upon correct, i.e. accurate, results that inform us of the true value of a particular analyte in a patient, whilst monitoring requires laboratory analytical methods to have a low imprecision, that is, a high degree of reproducibility from day to day, week to week or even month to month. Thus, reliable tests are essential if appropriate clinical decisions are to be made, but even the best available methods are subject to some analytical error from time to time. Furthermore, we would not expect the result for any given parameter to be exactly the same every time a sample is collected from the same person, because random fluctuations will occur due to, for example, feeding/eating cycles or patterns of exercise or degrees of stress that the individual experiences. Additionally, there are established biological rhythms that bring about more predictable changes in body chemistry. All such fluctuations can be described as normal biological variation.

Regular and frequent monitoring of plasma electrolyte (especially Na^+) concentrations is often required to assess the patient's change of hydration. Given that any laboratory result is subject to analytical error and change due to normal physiological intra-individual variation, the clinician needs to know whether a change in, say, plasma sodium

Table 2.6 Typical values for error

Measurement	Analytical error (mmol/L)	Individual biological variation (mmol/L)
Sodium	1.0	1.6
Potassium	0.15	0.2
Bicarbonate	0.5	1.5
Chloride	1.0	2.0

concentration represents a real change in the patient's status or is simply due to accumulated errors. Fortunately, the impact of combined analytical error and physiological variation on consecutive results can be assessed statistically.

$$\text{Significant change in concentration} = 2.8\sqrt{AE^2 + BV^2}$$

where

AE = analytical error and
BV = intra-individual biological (physiological) variation

2.8 is a statistical weighting factor giving 95% confidence in the calculated value.

Table 2.6 above gives examples of the magnitude of biological and analytical variations for common analytes.

Here is a worked example. Two consecutive results for plasma sodium concentration for a patient were 148 mmol/L and 145 mmol/L. Does this represent a real change or can it be ascribed simply to a combination of 'errors'?

$$AE = 1.0\,\text{mmol/L}$$
$$BV = 1.6\,\text{mmol/L}$$

$$\begin{aligned}\text{Significant change} &= 2.8 \times \sqrt{1^2+1.6^2}\\ &= 2.8 \times \sqrt{1+2.56}\\ &= 2.8 \times 1.9\\ &= 5.3\,\text{mmol/L}\end{aligned}$$

The observed difference in the two Na results is *less* than the critical difference of 5.3 mmol/L, so the decrease in concentration is *not* significant ($p = 0.05$) and the difference of 3 mmol/L could have arisen just by chance.

Table 2.7 Typical reference ranges (see also Appendix II)

Analyte	Reference range
Albumin	35–50 g/L
Bicarbonate	23–28 mmol/L
Calcium: total	2.25–2.55 mmol/L
ionised Ca^{2+}	1.15–1.30 mmol/L
Chloride	97–107 mmol/L
Magnesium	0.7–1.1 mmol/L
Potassium	3.5–4.5 mmol/L
Sodium	135–145 mmol/L
Urea	3.0–7.0 mmol/L

Population-based reference ranges, as shown in Table 2.7 and quoted throughout the text in connection with data interpretation exercises, should never be interpreted too strictly. Such ranges are derived by plotting the spread of results obtained from analysing samples from apparently healthy individuals. If the data distribution appears to be Gaussian (also called a 'normal distribution') the reference range includes 95% of the data points around the mean value, i.e. mean ± 2 standard deviations. Clearly, then, 5% of apparently healthy individuals fall outside the range and so could be classified as 'abnormal'. Furthermore, there may be differences in normal physiological values due to gender, age, ethnicity, and the time of day of sampling; the method of analysis may also have an impact on the perceived range. Thus, an apparently 'abnormal' result (just outside the reference range), especially if marginal, does not necessarily indicate the presence of pathology, nor does a normal value necessarily indicate absence of an underlying abnormality in the patient's condition. Clinicians need to consider the likelihood of a pathology relative to the laboratory results obtained. Only on certain occasions can data be used definitively to 'rule in' or 'rule out' a tentative diagnosis.

Finally, the clinician must be aware of drugs or medicines the patient may be taking, legitimately or not, as many can have either a genuine physiological effect which may complicate interpretation, or the drug, being a chemical, may have an interference effect on the actual analysis. The self-administration of 'herbal' or 'natural' remedies has attracted a lot of attention in this regard.

SECTION 2.xx

Sodium, protein and water

Recall from Section 1.xii that the plasma concentrations of Na^+ and protein are mainly responsible for maintaining osmotic balance (i.e. water distribution) between the ICF and the ECF. Although cell membranes are permeable to sodium and potassium, the action of the Na^+-K^+ ATP'ase pump maintains concentration gradients for both ions between the ICF and the ISF, thus ECF [Na^+] in particular is the main regulator of fluid distribution between the ICF and the ISF (which is of course part of the ECF). Clinically significant hypernatraemia (plasma [Na^+] greater than approximately 148 mmol/L) is associated with hyperosmolality, and hyponatraemia (plasma [Na^+] less than approximately 132 mmol/L) is associated with hypo-osmolality, thus significant changes in plasma sodium concentration can lead to either cellular dehydration or overhydration as water becomes redistributed between compartments.

An increase in the plasma [Na^+] will cause an increase in the ECF osmolality; this will draw water out of cells, effectively diluting the plasma. When the results of laboratory tests are examined, the rise in plasma sodium concentration may not appear significant because there will have been an increase in blood volume. Conversely, a fall in ECF [Na^+] will allow water to move from the ISF into the cells, and although blood volume and pressure will decrease and plasma [Na^+] will be low, the observed change may not always be significantly below the reference range limit of 135 mmol/L.

Pathological changes in plasma [Na^+] may therefore be 'masked' by simultaneous changes in plasma water volume, and laboratory results may be difficult to interpret because movements of sodium and

Essential Fluid, Electrolyte and pH Homeostasis, First Edition. Gillian Cockerill and Stephen Reed.

water across fluid compartment boundaries are invariably linked. For this reason, the abnormalities of water and sodium homeostasis are always discussed together. Changes in potassium homeostasis are often accompanied by acid-base abnormalities, as described in Part 3.

Plasma albumin concentration is the prime regulator of fluid movement between the blood vascular space and the ISF. A low plasma protein concentration causes oedema, which is the collection of fluid around cells, i.e. an increase in ISF volume.

As described in Section 2.vii, abnormalities leading to disruption of the balance in the forces which are involved in Starling's hypothesis may lead to excessive accumulation of fluid in tissue spaces. If the balance between these forces forcing fluid out of the vascular compartment and those tending to 'draw' fluid back into the plasma becomes compromised, ISF volume increases and oedema is the result. The increase in ISF volume is termed an exudate or a transudate according to the process that caused the accumulation.

Exudative oedema arises as part of an inflammatory reaction, and the fluid that accumulates is often protein (immunoglobulin) rich, whereas transudative oedema is due to leakage of essentially protein-poor fluid from the intravascular space consequent upon disruption of osmotic and haemodynamic forces controlling fluid movement from plasma to ISF. Specific causes of transudative oedema include heart failure, hepatic cirrhosis and nephrotic syndrome. Clearly irrespective of the cause, there is an increase in total ECF volume, but all three examples are associated with some degree of sodium retention by the kidney and therefore expansion of the intravascular volume, i.e. plasma and ISF volumes both rise.

The heavy proteinuria associated with nephrotic syndrome is simply a symptom of many possible underlying renal pathologies affecting the integrity of the renal glomerulus. The characteristic finding in nephrotic syndrome is heavy proteinuria, often exceeding 5 g/day (normal urinary protein loss is ~150 mg/day) as the glomeruli become functionally 'leaky'. If protein loss exceeds the rate of albumin synthesis in the liver, the patient quickly becomes hypoproteinaemic (hypoalbuminaemic), so reducing the plasma oncotic pressure. Initially, the pooling of plasma fluid as oedema fluid leads to a fall in intravascular volume. The renin/aldosterone/ADH response mechanism operates to Figure 2.11, Section 2.vii and plasma volume increases, contributing further to the hypoalbuminaemia. Oedema formation continues and the process repeats itself. Thus, sodium retention is seen to be a normal response to a physiological stimulus, although it has also been suggested that there may also be disordered Na^+ handling in the distal tubule leading to uncontrolled sodium reabsorption.

Oedema associated with cirrhosis is also associated with hypoalbuminaemia, but this time due to reduced synthesis rather than increased loss. The situation is aggravated by portal hypertension arising from fibrosis-induced disruption to blood flow through the liver. Hypertension increases blood hydrostatic pressure, which accompanied by the low plasma albumin concentration leads to oedema (ascites) formation in the peritoneal space. Sodium reabsorption in the nephron is stimulated by the reduced plasma volume. An alternative view of ascites formation is that Na^+ retention is triggered by an autonomic 'hepato-renal' reflex, directly modifying the actions of the nephron.

Heart failure, more usually called congestive cardiac failure, CCF, means that the pump effect of the heart is so poor that blood flow to and from the tissues is reduced. Gas exchange in the lungs is compromised and patients often feel 'short of breath' arising from pulmonary congestion. As with the nephrotic syndrome, there is a perceived volume depletion due to reduced arterial flow, and the kidney responds by producing renin. Ultimately, therefore, Na^+ retention occurs.

Treatment for oedema relies heavily upon the use of diuretics. There are many such drugs, but all can be classified as either (a) those that directly affect segments of the nephron and in particular target sodium handling, or (b) drugs that modify the neurohormonal control of sodium reclamation in the nephron.

Normal ion and/or water handling by regions of the nephron can be modified by carbonic anhydrase inhibitors (e.g. acetazolamide), thiazides acting on Na^+ reabsorption in the distal regions of the tubule, and so-called 'loop diuretics' (e.g. furosemide) which inhibit the Na-K-Cl transporter in the thick ascending limb of the loop of Henle. One common and unwanted effect of many sodium-modifying diuretics is excessive loss of potassium, sufficient to cause hypokalaemia in some cases. However, the group of drugs known collectively as 'potassium sparing' diuretics target the principal cells of the collecting duct. Spironolactone is a mineralocorticoid receptor antagonist effectively making the cells less responsive to aldosterone, whereas amiloride specifically inhibits ENaC sodium transporter without affecting K^+-channels in the apical membrane of the collecting duct cells.

Of the drugs that interrupt the aldosterone/ADH axis, the most well known is probably the group known as angiotensin converting enzyme (ACE) inhibitors, such as captopril and ramipril. ACE inhibitors also have the advantage of preventing the proteolysis of bradykinin (also a substrate for ACE). Bradykinin is a vasodilator, an action that complements the antihypertensive effects brought about by reduction in angiotensin II production. Losartan and valsartan exert their anti-hypertensive effects as antagonists of AT_1 receptors, which mediate the normal physiological actions of angiotensin.

In contrast to the drugs discussed above, nesiritide is a receptor agonist rather than an antagonist. One of the relatively new drugs which are engineered recombinant peptides, nesiritide is an analogue of B-type natriuretic peptide.

Hyper- and hyponatraemia arise due to the *relative* changes in $[Na^+]$ and water content of body fluids, and the terms do not necessarily indicate an absolute deficiency or excess total body sodium. Most cases of disturbances in sodium and water balance are the result of excessive sodium and/or water loss from the body via the gut or the kidney. Alternatively, abnormalities arise due to changes in the secretion or action at their target sites of regulatory hormones, ADH and aldosterone, leading to either loss or retention of sodium or water or more likely both. Some common causes of hyper-and hyponatraemia are listed in Table 2.8.

The physiological consequences of hyper- and hyponatraemia may be severe, especially in the brain which is subject to shrinkage (hypernatraemia) and swelling (hyponatraemia). It has been suggested that the brain is able to produce its own 'osmolality buffers' called 'idiogenic osmols', thought to be amino acids (such as taurine) and polyols to counteract the chronic effects of hypernatraemia. The increase in cerebral intracellular tonicity generated by these compounds, plus the physiological redistribution of sodium, allows the brain to recover from dehydration. However, the compensatory synthesis of these compounds takes a day or two to be effective so there is no benefit in conditions causing an acute hypernatraemia.

The symptoms observed in hypernatraemia range from mild (nausea, thirst, lethargy and irritability) to moderate (muscle weakness), and severe effects on the CNS (fits, confusion and even coma). The severity of any symptoms may be due to the rate of change of the analyte concentration and not simply to the absolute value. The mortality associated with cases of hypernatraemia has been estimated to be

Table 2.8 Abnormalities in plasma sodium concentration

	Physiological outcome	
Caus	Hypernatraemia	Hyponatraemia
Change in body fluid volume	Loss of hypotonic fluid, e.g. diarrhoea and vomiting; excessive sweating; osmotic diuresis	Chronic water overload: fluid intake outpaces the ability of the kidney to excrete the water load, e.g. renal failure
	Pure water depletion, e.g. inadequate water intake	Inappropriate secretion of ADH (*SIADH* = *syndrome of inappropriate ADH secretion*) which may be ectopic or eutopic. Some drugs increase ADH secretion, others increase the sensitivity of the distal nephron to the action of ADH
Endocrine causes	Water loss: diabetes insipidus (the mechanism is loss of hypotonic fluid as above)	Water gain: SIADH (above)
	Salt gain, e.g. due to excess cortisol or excess aldosterone	Salt loss: Addison's disease
Other causes	Iatrogenic ('caused by therapy') salt gain, e.g. infusion of Na-rich i.v. fluids	Congestive cardiac failure (CCF), liver cirrhosis and nephrotic syndrome Sick cell syndrome
	Accidental or deliberate salt ingestion.	Artefactual (pseudohyponatraemia; inappropriate sample collection or storage)

greater than 50%, especially if the rise in plasma sodium concentration occurs rapidly.

Hyponatraemia is one of the most commonly encountered abnormalities of fluid and pH homeostasis. The CNS symptoms of severe hyponatraemia are similar to those of hypernatraemia: confusion, convulsions, and muscle weakness. Symptoms such as these develop if plasma sodium concentration falls rapidly to less than about 125 mmol/L, but may not be apparent until the sodium is less than 120 mmol/L if the decline is gradual. Clinically, patients lose skin elasticity (a pinched skin fold does not recoil quickly) and have sunken eyes. Severe hyponatraemia is a medical emergency that carries a high mortality risk. A low plasma sodium concentration is most commonly the result of water excess causing a dilutional hyponatraemia.

SECTION 2.xxi

Hypernatraemia

Signs and symptoms of hypernatraemia are often vague and reflect the underlying cause (fluid loss or salt gain), but include:

- Dry mucous membranes
- Decreased tissue turgor
- Thirst
- Disorientation
- Confusion
- Muscle weakness
- Convulsions
- Coma

For the concentration of sodium in ECF to rise, there must be either a decrease in volume or an increase in the quantity of sodium present in the fluids. Salt gain is associated with unintentional salt intake, rarely via the oral route but more likely as Na-rich intravenous fluids as in the treatment of acidosis with sodium bicarbonate. Alternatively, hypernatraemia may follow reduced fluid intake or excessive loss of Na-poor fluids, in which case there will not be an increase in total body sodium.

In most cases of hypernatraemia there is loss of both sodium and water, but the loss of water predominates over the loss of salt and the plasma osmolality rises. The fluid redistribution between ICF and ECF that follows any loss of salt-containing fluid depletes the ECF (plasma volume in particular) proportionately more than does the loss of an equivalent volume of pure water. Consequently, there may be significant hypovolaemia but a less marked rise in plasma osmolality.

Essential Fluid, Electrolyte and pH Homeostasis, First Edition. Gillian Cockerill and Stephen Reed.

Loss of pure water results in depletion of both ICF and ECF in proportion to their initial volumes. Importantly, the plasma loss is approximately 3.2/42 of the total fluid lost (recall plasma volume = 3.2 litres; total body water = 42 litres) and severe hypovolaemia is uncommon. Plasma osmolality may, however, be significantly raised; the *amount* of sodium in the plasma has not changed but it is present in a reduced volume.

Most cases of water depletion are due to loss of hypotonic fluid, that is, loss of water in excess of electrolytes rather than loss of water alone. Thus, loss of pure water from the body is fairly rare, arising only if water is unavailable or the person is unable to drink due, for example, to a mechanical obstruction or unconsciousness.

Hypernatraemia due to loss of fluid through renal or non-renal routes

Excessive renal losses will arise in circumstances where there is elimination of large volumes of dilute urine, which can be viewed as a form of pure water loss. For example, in health, a typical 24-hour urine volume is approximately 1500 mL containing, say, 100 mmol of sodium, but if the urine volume rises to 4000 mL containing the same quantity of sodium, we could say that in effect 2.5 litres of 'salt-free' water have been excreted. Such excessively large volumes of urine may be lost if there is a functional deficiency of anti-diuretic hormone, a condition known as diabetes insipidus (DI). If blood volume and pressure fall for any reason and appropriate homeostatic mechanisms (Section 2.iii) operate normally, ADH production by the pituitary increases and the lost volume is restored by increased reabsorption of water in the nephron. Inadequate secretion of ADH from the pituitary (central or neurogenic DI) or failure of the distal nephron cells to respond to ADH (nephrogenic DI) means that the individual excretes large volumes (perhaps as much as 10 litres a day) of dilute urine. A similar, but less severe diuresis occurs when the glomerular filtrate contains an abnormally high amount of urea or glucose. This osmotic diuresis may occur in diabetic patients during hyperosmotic coma.

In cases where the fluid loss continues or the ADH response is inadequate, initial treatment of pure water loss is aimed at restoring the lost fluid, so hypotonic saline or glucose/hypotonic saline solution may be infused. In contrast, the first priority in a patient with hypovolaemia is to restore blood volume to prevent circulatory collapse and possible organ (especially renal) damage. Diabetes insipidus may be treated by the use of ADH analogues.

Extrarenal losses of salt-poor water may occur due to fever or in thyrotoxicosis. The hyperventilation associated with a fever may contribute to hypernatraemia, but sweating will cause some salt, as well as water, loss.

Hypernatraemia due to salt gain

Increased sodium loading will lead to hypernatraemia. Deliberate ingestion of quantities of table salt is uncommon (and indeed may be unpalatable), but salty foods taken over a long period of time raise the blood volumes and therefore blood pressure, with serious effects on organs especially the blood vascular system. Hypertension is a risk factor for cardiovascular diseases such as myocardial infarction or stroke. Accidental overload with sodium may be due to inappropriate use of intravenous infusions (e.g. over-correction of hyponatraemia) or excessive use of sodium bicarbonate to treat acute acidosis.

Given the importance of hormonal regulation of sodium and water balance, endocrine dysfunction can lead to hypernatraemia. Specifically, disorders of the adrenal cortex will result in hypernatraemia due to salt gain in, for example, Cushing's disease or Conn's syndrome, where there is an inappropriate concentration of mineralocorticoid due to increased reclamation of filtered Na^+ from the glomerular filtrate.

Conn's syndrome (primary hyperaldosteronism) arises due to adrenal hyperplasia or adrenal adenoma. Sodium retention is often masked by an increase in blood volume (hypertension), and a low plasma potassium concentration with kaliuria (>30 mmol K^+ lost in the urine per day) may be the main presenting feature.

An elevation of plasma cortisol may be due to adrenal disease, excessive production of pituitary adrenocorticotrophic hormone (ACTH) or ectopic ACTH secretion. Normally cortisol is only weakly active in causing sodium retention due to the action of the inactivating enzyme 11-β-hydroxysteroid dehydrogenase in the nephron, but in cases of hypercortisolaemia the enzyme becomes swamped, leading to sodium retention and hypertension. As with Conn's, the plasma [Na^+] may not appear very high due to the dilution effect of increased water reabsorption.

Case studies

(1) Female Age: 85 yrs A&E admission.
Found unconscious at home.
Rehydration fluids were infused on admission.

	On admission	24 hrs later	48 hrs later	Ref.
Plasma				
Na^+	163 mmol/L	152 mmol/L	143 mmol/L	135–145 mmol/L
K^+	3.4 mmol/L	3.2 mmol/L	3.5 mmol/L	3.5–4.5 mmol/L
Urea	29.5 mmol/L	18.1 mmol/L	9.9 mmol/L	3.0–7.0 mmol/L
Creatinine	180 μmol/L	125 μmol/L	105 μmol/L	80–120 μmol/L
HCO_3^-	28 mmol/L	27 mmol/L	28 mmol/L	23–28 mmol/L
Osmolality	395 mmol/kg	335 mmol/kg	300 mmol/kg	285–295 mmol/kg
Urine				
Na^+	80 mmol/L		30 mmol/L	<100 mmol/L
Osmolality	760 mmol/kg		380 mmol/kg	

Interpretation: dehydration due to reduced fluid intake. This lady had presumably been lying unattended for some time. The high osmolality of the initial urine sample indicates the sample was very concentrated because the kidneys were attempting to reabsorb water. Following rehydration, both the plasma and urine are less concentrated.

(2) Female Age: 23 yrs A&E admission, following fall down stairs.
Head injury.

	On admission	24 hrs later	Ref.
Plasma			
Na^+	138 mmol/L	160 mmol/L	135–145 mmol/L
K^+	4.0 mmol/L	3.8 mmol/L	3.5–4.5 mmol/L
Urea	4.5 mmol/L	6.6 mmol/L	3.0–7.0 mmol/L
Creatinine	95 μmol/L	115 μmol/L	80–120 μmol/L
HCO_3^-	24 mmol/L	25 mmol/L	23–28 mmol/L
Osmolality		335 mmol/kg	285–295 mmol/kg
Urine			
Na^+		<5 mmol/L	<100 mmol/L
Osmolality		155 mmol/kg	
24-hr volume:		4.5 litres	

Interpretation: traumatic diabetes insipidus, ADH deficiency. This patient is losing large volumes of sodium-poor and very dilute fluid via the kidneys and is therefore dehydrated.

(3) Male Age: 10 yrs Oedematous appearance

	On admission	Ref.
Plasma		
Na^+	148 mmol/L	135–145 mmol/L
K^+	3.0 mmol/L	3.5–4.5 mmol/L
Albumin	29 g/L	35–50 g/L

Interpretation: nephrotic syndrome. Loss of albumin via the kidney results in a fall of plasma oncotic pressure. Oedema occurs and the normal physiological monitoring mechanisms respond to a fall in blood pressure/volume. This leads to aldosterone-induced Na^+ and water retention but potassium loss.

(4) Female Age: 34 yrs Known diabetic.
Admitted in a semi-coma.
Low BP and tachycardia and clinically dehydrated.

	On admission	Ref.
Plasma		
Na^+	160 mmol/L	135–145 mmol/L
K^+	4.6 mmol/L	3.5–4.5 mmol/L
Urea	14.5 mmol/L	3.0–7.0 mmol/L
HCO_3^-	22 mmol/L	23–28 mmol/L
Glucose	41.6 mmol/L	3.5–5.5 mmol/L fasting 3.5–11 mmol/L non-fasting
Osmolality	384 mmol/kg	285–295 mmol/kg
Urine		
Glucose	Present +++	
Osmolality	265 mmol/kg	

Interpretation: This is hyperosmotic diabetic coma. The severe glycosuria has led to an osmotic diuresis and the patient is losing large volumes of fluid via the kidneys.

(5) Female Age: 61 yrs Presented to her GP complaining of fatigue and weakness. She has been a heavy smoker for 40 years
Skin pigmentation noted.

	On admission	Ref.
Plasma		
Na^+	149 mmol/L	135–145 mmol/L
K^+	2.4 mmol/L	3.5–4.5 mmol/L
Urea	5.9 mmol/L	3.0–7.0 mmol/L
Cl^-	92 mmol/L	97–107 mmol/L
HCO_3^-	>45 mmol/L	23–28 mmol/L
Osmolality	299 mmol/kg	285–295 mmol/kg
On referral to endocrinologist at hospital		
Na^+	152 mmol/L	135–145 mmol/L
K^+	2.3 mmol/L	3.5–4.5 mmol/L
9 am cortisol	1850 nmol/L	150–650 nmol/L

Interpretation: Cushing's syndrome due to ectopic ACTH secretion from the bronchus. Although cortisol is only a weak mineralocorticoid, the grossly elevated concentration is exerting an aldosterone-like effect. The effect on plasma potassium is more noticeable in this case.

(6) Male Age: 70 yrs On intravenous infusion to correct dehydration

	On admission	Ref.
Plasma		
Na^+	153 mmol/L	135–145 mmol/L
K^+	2.9 mmol/L	3.5–4.5 mmol/L
Urea	2.3 mmol/L	3.0–7.0 mmol/L
Creatinine	45 μmol/L	80–120 μmol/L
Glucose	2.8 mmol/L	3.5–5.5 mmol/L fasting 3.5–11 mmol/L non-fasting

Interpretation: Sample taken from drip arm. Urea and creatinine are unusually low for a patient of this age, suggesting that most of the plasma constituents have been diluted by the infusion of 'normal' 0.9% saline.

SECTION 2.xxii

Hyponatraemia

Signs and symptoms

As with hypernatraemia, the specific clinical signs often reflect the underlying cause of which hyponatraemia is itself just a symptom. For example:

- Generalised dehydration; dry skin, thirst
- slow capillary refill if there is loss of Na-rich fluid
- Fatigue and lethargy
- Irritability
- Weakness
- Nausea
- Headache
- Confusion, delirium, fitting possibly leading to coma

Hyponatraemia is the most commonly met electrolyte imbalance in routine practice. Two broad categories of underlying cause are (i) water excess or (ii) sodium depletion. In the case of water excess there will be evidence of generalised dilution of blood (e.g. hypoalbuminaemia) and a low packed cell volume (PCV, haematocrit). Paradoxically, hypoalbuminaemia may itself be associated with hyponatraemia as a consequence of oedema. Fluid collection in the tissue spaces means that the return volume of blood reaching the heart is reduced, and this prompts an appropriate ADH response which expands the plasma volume but compounds the hyponatraemia. In addition, albumin carries a net negative charge at normal blood pH, so hypoalbuminaemia represents a reduction in 'total anions'. In order to maintain electrical neutrality there must be an equal reduction in 'total cations', and sodium is the major cation. Primary sodium depletion is rare and usually only seen when loss of isotonic fluids is replaced with water, as in the case of excessive sweating. Excessive loss of sodium can

Essential Fluid, Electrolyte and pH Homeostasis, First Edition. Gillian Cockerill and Stephen Reed.

occur via the kidney (Addison's disease, see below, or diuretic drugs) or the gut. Measurement of urine sodium concentration is a useful determinant of the cause; if urine [Na^+] is less than 10 mmol/L, a non-renal cause is indicated.

Two endocrine-related causes are important to note. Firstly, in Addison's disease the ability of the adrenal cortex to produces steroid hormones, in particular aldosterone, is compromised. Secondly, there may be overproduction of anti-diuretic hormone from the hypothalamic-pituitary axis or ADH secretion from an ectopic site, usually a tumour. The 'syndrome of inappropriate ADH secretion' (SIADH) causes a dilutional hyponatraemia. SIADH is not a specific disease as such but merely a consequence of a number of pathologies including malignant tumours, CNS disorders (infection, head injury), drugs (narcotics, hypnotics, anti-convulsants and anti-neoplastics) and pulmonary disease. Notably, one of the body's responses to injury, including surgery, is to increase ADH production, so hyponatraemia is not uncommon in 'post-op' patients.

A low plasma [Na^+] may occur when the total body water (TBW) is increased (i.e. essentially a dilution of normal total body Na in a larger volume), *or* if TBW is decreased (hypovolaemia with a greater degree of Na loss), *or* if TBW is normal (euvolaemia). All situations will bring about hypotonicity of the ECF. If a sample shows hyponatraemia with a normal osmolality, pseudohyponatraemia due to hyperlipidaemia or hyperproteinaemia must be suspected, whereas hyponatraemia with hyperosmolality usually indicates fluid redistribution from ICF to ECF during hyperglycaemia.

Physiologically and pathologically, it may be difficult to separate changes in sodium concentration from changes in blood volume. Hypovolaemia associated with haemorrhage, diarrhoea and vomiting, inappropriate use of diuretics or 'pooling', as in pleural or peritoneal effusions, will cause compromised blood flow and therefore oxygenation of tissues. In general terms hypovolaemia can be classified as:

- Isotonic hypovolaemia: loss of water and electrolytes in approximately equal amounts so plasma chemistry may not show any dramatic changes;
- Hyponatraemic hypovolaemia: water and salt are lost together but only water is replaced;
- Hypernatraemic hypovolaemia: disproportionate loss of water over sodium.

Case studies

(1) Female Age: 56 yrs i.v. fluids post abdominal surgery.

	On admission	Ref.
Plasma		
Na^+	129 mmol/L	135–145 mmol/L
K^+	3.2 mmol/L	3.5–4.5 mmol/L
Urea	2.5 mmol/L	3.0–7.0 mmol/L
Creatinine	85 μmol/L	80–120 μmol/L
Glucose	18.5 mmol/L	3.5–5.5 mmol/L fasting 3.5–11 mmol/L non-fasting

Interpretation: The infusion used in this case was half-normal saline with 2.5% glucose. Blood was taken from the drip arm so the glucose concentration is erroneously high but other constituents are diluted by the fluid. Compare this with case 6 on page 178.

(2) Male Age: 76 yrs 36 hrs post-operative.

	On admission	Ref.
Plasma		
Na^+	127 mmol/L	135–145 mmol/L
K^+	4.1 mmol/L	3.5–4.5 mmol/L
Urea	4.3 mmol/L	3.0–7.0 mmol/L
Creatinine	65 μmol/L	80–120 μmol/L
Osmolality	270 mmol/kg	285–295 mmol/kg
Urine		
Na^+	$<$10 mmol/L	
Osmolality	495 mmol/kg	

Interpretation: SIADH due to post-op trauma (also called surgical hyponatraemia). Physiological stress of surgery causes a cortisol response and inappropriate secretion of aldosterone and ADH, all of which are likely to affect sodium and water balance resulting in salt and water retention. The low concentration of sodium in the urine shows that most of the filtered load is being reabsorbed and the osmolality indicates the excretion of concentrated urine. The net effect of ADH is greater than aldosterone and the small mineralocorticoid action of cortisol, so the plasma [Na^+] is low.

(3) Male Age: 73 yrs Heavy smoker for over 50 years

	On admission	Ref.
Plasma		
Na^+	116 mmol/L	135–145 mmol/L
K^+	3.5 mmol/L	3.5–4.5 mmol/L
Urea	6.0 mmol/L	3.0–7.0 mmol/L
Creatinine	108 μmol/L	80–120 μmol/L
Osmolality	255 mmol/kg	285–295 mmol/kg
Urine		
Na^+	68 mmol/L	
Osmolality	300 mmol/kg	

Interpretation: The diagnosis for this patient was small cell lung cancer with ectopic secretion of ADH. The relatively high concentration of sodium in the urine seems paradoxical but is due to volume expansion of the ECF, probably triggering release of natriuretic factors.

(4) Female Age: 57 yrs Architect.

Recent history: frequent headache, difficulty in focusing vision and some double vision (has worn spectacles since childhood). There have been two periods of vomiting in the last week; feelings of disorientation.

Low plasma [Na^+] and low plasma osmolality. High C-reactive protein (CRP) and erythrocyte sedimentation rate (ESR). All else normal.

Follow-up: thyroid hormones and cortisol all low. Slightly reduced response to short synacthen stimulation confirming that the defect lies in the pituitary not the adrenal cortex itself

Interpretation: Non-functioning pituitary adenoma which has destroyed the normal tissue resulting in hormone deficiency. The blurred vision is typical of pituitary tumours because the gland lies close to the optic chiasma, a part of the brain involved with visual signal processing. The C-reactive protein and erythrocyte sedimentation rate results are non-specific, indicating simply that there is an underlying inflammatory condition in this patient.

(5) Male Age: 70 yrs Taken to A&E by a neighbour who was aware of the patient's diabetes, noticed the patient acting in a 'confused and disoriented' fashion. Patient records showed assessments for Alzheimer's disease had also been carried out. On examination, he was clearly dehydrated and urine analysis showed a 4+ positive reaction for glucose.

	On admission	Ref.
Plasma		
Na^+	125 mmol/L	135–145 mmol/L
K^+	3.7 mmol/L	3.5–4.5 mmol/L
Bicarbonate	20 mmol/L	23–28 mmol/L
Urea	13.8 mmol/L	3.0–7.0 mmol/L
Glucose	35.7 mmol/L	3.5–11 mmol/L non-fasting
Osmolality	320 mmol/kg	285–295 mmol/kg

Interpretation: The marked hyperglycaemia in this patient has caused movement of water out of cells, and an osmotic diuresis due to glycosuria will account for the dehydration and confusion. Although there is obvious hyponatraemia, the osmolality is raised due to the severe hyperglycaemia; the calculated osmolarity is 298 mmol/kg, so the osmol gap is a little higher than would be expected. There are only few instances of hypertonicity with a low plasma sodium concentration.

SECTION 2.xxiii

Disturbances of potassium homeostasis

Disturbances in potassium homeostasis are usually viewed as arising from either renal or non-renal causes, i.e. failure of normal homeostatic mechanisms. With the exception of iatrogenic causes, increased or decreased intake of potassium is rare. Some common causes of abnormal plasma potassium concentration are shown in Sections 2.xxiv and 2.xxv. It must be appreciated that plasma potassium concentration is a poor indicator of total body potassium because of the predominantly intracellular location of the ion.

Hypokalaemia ($[K^+] < 3.3$ mmol/L) is more commonly seen in clinical practice than hyperkalaemia ($[K^+] > 4.8$ mmol/L), but both may lead to muscle weakness and indeed the major clinical concern is cardiac arrhythmia. Characteristic ECG changes appearing when taller than normal T waves are seen as the $[K^+]$ rises above approximately 6.5 mmol/L, or lowered T wave peaks and the appearance of a U wave in hypokalaemia.

Changes in potassium balance often reflect in acid-base abnormalities. This is partly due to redistribution of H^+ and K^+ between ICF and ECF, but is also due to effects within the renal cells. A normal physiological response is for H^+ to move into cells, so to maintain electrical neutrality across the membrane, K^+ leaves the cell. Similar movements in renal tubular cells affect potassium excretion; alkalosis leads to kaliuresis (excess K^+ lost in urine) and acidosis results in potassium retention in the plasma. Also, acidosis promotes Na^+-H^+ exchange, rather then Na^+-K^+ exchange, in the nephron, further contributing to potassium retention. Conversely, hyperkalaemia contributes to metabolic (non-respiratory) acidosis by inhibiting the

Essential Fluid, Electrolyte and pH Homeostasis, First Edition. Gillian Cockerill and Stephen Reed.

production of ammonia in the tubular cells, so net acid excretion as NH_4^+ is reduced.

Abnormal effects seen in alkalosis include peripheral vasodilation (leading to reduced blood pressure and thus reduced glomerular filtration rate, with consequent effects on NaCl and water handling), increased ammoniagenesis and therefore net acid excretion. Hyperglycaemia (elevated blood glucose concentration) may occur as insulin release is impaired by hypokalaemia. The physiological link between potassium and insulin goes beyond secretion of the hormone; insulin and glucose injections are used to treat hyperkalaemia because entry of potassium into cells is facilitated by insulin.

SECTION 2.xxiv

Hyperkalaemia

The ability of the healthy kidney to excrete potassium means that hyperkalaemia without underlying renal or adrenal disease is rare.

Signs and symptoms

- Cardiac arrhythmia; fibrillation and cardiac arrest when $[K^+]$ >10 mmol/L
- Abnormal ECG; peaked T waves when $[K^+]$ ~7 mmol/L, prolonged QRS complex when $[K^+]$ ~9 mmol/L
- Muscle weakness starting often in the lower limbs, progressing to paralysis if prolonged
- Numbness or tingling sensation in the fingers
- Gastrointestinal signs such as nausea and diarrhoea.

Non-renal causes

Redistribution between ICF and ECF may be the result of:

- Tissue necrosis: K^+ leaks out of cells following, for example, crush injury, major surgery, or rhabdomyolysis (a condition involving muscle tissue degradation).
- Impaired Na^+-K^+ ATP'ase activity, which may be prompted by reduced aldosterone concentration or inhibitory effects of drugs such as digoxin. Sick cell syndrome also causes impairment of the sodium-potassium pump.
- Acidosis: K^+ moves out of cells in exchange for protons which enter cells to be buffered.
- Lack of insulin: insulin promotes the cellular uptake not only of glucose but also potassium, phosphate and magnesium.

Essential Fluid, Electrolyte and pH Homeostasis, First Edition. Gillian Cockerill and Stephen Reed.

- Infusions: K^+-rich i.v. fluids; massive blood transfusion (blood stored even at 4°C consumes glucose and the action of the Na^+-K^+ pump becomes compromised).

Renal-related causes

- Acute renal failure.
- Reduced sodium delivery to the distal tubule (hyponatraemia) so there is reduced Na^+/K^+ exchange.
- Reduced aldosterone secretion (Addison's disease) or impaired aldosterone action on nephron. Hyporeninaemic hypoaldosteronism (type 4 renal tubular acidosis)
- Drugs: amiloride, spironolactone (both are diuretics); non-steroidal anti-inflammatory drugs (NSAIDs); cyclosporin (an immunosuppressive often used post-organ transplant).

Case studies

(1) Female Age: 40 yrs A&E admission. ?drug ingestion several hours previously. ECG shows an abnormal pattern.

	On admission	Ref.
Plasma		
Na^+	133 mmol/L	135–145 mmol/L
K^+	6.4 mmol/L	3.5–4.5 mmol/L
Urea	39.3 mmol/L	3.0–7.0 mmol/L
Creatinine	1725 μmol/L	80–120 μmol/L
HCO_3^-	18 mmol/L	23–28 mmol/L
Glucose	5.6 mmol/L	3.5–5.5 mmol/L fasting 3.5–11 mmol/L non-fasting

Interpretation: These data are consistent with acute nephrotoxicity. The drug taken by this patient seems to have caused serious renal damage and it appears that she is in acute renal failure.

(2) Female Age: 31 yrs She complains of feeling tired, weight loss and dizzy spells. Clinical examination shows hypotension and skin pigmentation.

	On admission	Ref.
Plasma		
Na^+	132 mmol/L	135–145 mmol/L
K^+	5.4 mmol/L	3.5–4.5 mmol/L
Urea	7.7 mmol/L	3.0–7.0 mmol/L
Basal cortisol	160 nmol/L	180–550 nmol/L
Basal ACTH	355 ng/L	10–50 ng/L
Cortisol post synacthen*	185 nmol/L	>500 nmol/L
Anti-adrenal antibodies present		

*Synacthen is a synthetic ACTH used to stimulate the adrenal cortex.

Interpretation: Addison's disease. Autoimmune destruction of the adrenal cortex leading to mineralocorticoid deficiency.

(3) Male Age: 26 yrs A known diabetic on insulin injections, he attended the local A&E whilst staying away from home with friends. Recently he had had a 'viral illness' and he noticed his blood glucose concentrations were not as well controlled as usual. He had experienced some nausea and polyuria and clinically he was dehydrated.

	On admission	Ref.
Plasma		
Na^+	147 mmol/L	135–145 mmol/L
K^+	6.6 mmol/L	3.5–4.5 mmol/L
Urea	6.8 mmol/L	3.0–7.0 mmol/L
Creatinine	115 μmol/L	80–120 μmol/L
HCO_3^-	20 mmol/L	23–28 mmol/L
Glucose	19.6 mmol/L	3.5–5.5 mmol/L fasting 3.5–11 mmol/L non-fasting

Interpretation: There is marked hyperglycaemia and hyperkalaemia. The urea and creatinine results, although within their respective reference ranges, are near the upper limits suggesting dehydration, probably due to a diuresis caused by the osmotic effect of glucose in urine. The hyperkalaemia is a reflection of acidosis arising from production of 'ketone bodies' from the metabolism of lipids. Hyperkalaemia and 'hyperprotonaemia' (acidosis) invariably co-exist as K^+ and H^+ move between ICF and ECF.

SECTION 2.xxv

Hypokalaemia

Hypokalaemia is a more common finding in routine practice than is hyperkalaemia. Given the distribution of potassium to the ICF, it must be remembered that low plasma potassium does not necessarily reflect potassium deficiency.

Signs and symptoms

- Cardiac dysrhythmia
- Muscle weakness
- Paralysis
- Irritability
- Confusion and disorientation perhaps leading to coma

Note the similarity of these symptoms with those of hyperkalaemia.

It is convenient to subdivide the main causes of hypokalaemia into renal and non-renal mechanisms.

Non-renal mechanisms

In these cases, urinary [K] is typically <25 mmol/L.

(a) Redistribution of K into cells.
- Alkalosis: K moves into cells as protons exit to help correct the high pH
- Adrenaline: β-adrenergic stimulation
  - Increased adrenergic stimulation (adrenaline surge)
  - Catecholamine secreting tumours (phaechromocytoma)
- Insulin: insulinoma (insulin-secreting tumour) or administration of insulin for treatment of diabetic emergency
- Hypercorticalism (Cushing's or increased renin secretion): this is usually associated with hypernatraemia as well as hypokalaemia. In Cushing's, the prevailing cortisol concentration

Essential Fluid, Electrolyte and pH Homeostasis, First Edition. Gillian Cockerill and Stephen Reed.

exceeds the capacity of 11-β-hydroxy steroid dehydrogenase to convert cortisol to cortisone, so the normally weak aldosterone-like mineralocorticoid effect of cortisol is not suppressed and potassium excretion is promoted (see Section 2.vi). Glycyrrhizinic acid, a component of liquorice, inhibits 11-β-hydroxy steroid dehydrogenase and brings about a similar hypokalaemic effect.

(b) Increased gastrointestinal losses due to diarrhoea and/or vomiting; purgative abuse; intestinal surgery, e.g. ileostomy. Note: gastric juice is K-poor but vomiting causes fluid depletion, which in turn stimulates an aldosterone response so the actual mechanism of K loss is urinary. Colonic fluid, however, is relatively K-rich, so in diarrhoea the loss is via the faecal route.

Renal-related causes of hypokalaemia

In these cases, increased losses, but [K] is typically >25 mmol/L.

- Diuretic therapy, e.g. thiazides; this is the commonest cause of hypokalaemia. Also osmotic diuresis
- Acute tubular necrosis (diuretic phase)
- Renal tubular acidosis types 1 and 2
- Drug nephrotoxicity, e.g. aminoglycosides.

Case studies

(1) Male Age: 65 yrs Known congestive cardiac failure (CCF); on treatment

	On admission	Ref.
Plasma		
Na^+	132 mmol/L	135–145 mmol/L
K^+	2.9 mmol/L	3.5–4.5 mmol/L
Urea	10.8 mmol/L	3.0–7.0 mmol/L
Creatinine	145 μmol/L	80–120 μmol/L

Interpretation: CCF can lead to fluid accumulation so patients are prescribed diuretics to help excretion of water. This is a case of diuretic-induced K^+ loss. Patients on certain diuretics should have their plasma potassium measured regularly.

(2) Male Age: 49 yrs GP referral. Patient complained of weakness and tiredness. Clinically, his BP was elevated.

	On admission	Ref.
Plasma		
Na^+	145 mmol/L	135–145 mmol/L
K^+	2.7 mmol/L	3.5–4.5 mmol/L
Urea	3.3 mmol/L	3.0–7.0 mmol/L
Creatinine	70 μmol/L	80–120 μmol/L
Osmolality	270 mmol/kg	285–295 mmol/kg
HCO_3^-	34 mmol/L	23–28 mmol/L
Cl^-	101 mmol/L	97–107 mmol/L
Urine		
Na^+	54 mmol/L	80–150 mmol/L
K^+	88 mmol/L	50–100 mmol/L
Creatinine	9.5 mmol/L	7.5–12 mmol/L

Interpretation: This is Conn's syndrome (primary hyperaldosteronism). The excess production of aldosterone has caused loss of potassium via the urine. The plasma sodium is at the upper limit, suggesting there may be some fluid overload. Although the 24-hour excretion of potassium is within the reference range, it is inappropriately high for someone with a low plasma [K^+] when values of less than 30 mmol/day are more typical. The bicarbonate concentration is high due to potassium loss (alkalosis and hypokalaemia invariably coincide).

(3) Female Age: 61 yrs Weakness, weight loss. HT skin pigmentation. X-ray shows a shadow in right lung.

	On admission	Ref.
Plasma		
Na^+	144 mmol/L	135–145 mmol/L
K^+	2.4 mmol/L	3.5–4.5 mmol/L
Urea	7.6 mmol/L	3.0–7.0 mmol/L
HCO_3^-	32 mmol/L	23–28 mmol/L
9am cortisol	650 nmol/L	150–650 nmol/L
Midnight cortisol	455 nmol/L	50–200 nmol/L
Plasma cortisol did not suppress with dexamethasone		

Interpretation: Cushing's syndrome (excess cortisol production) due to ectopic ACTH from lung tumour. Over-stimulation of the aldosterone-responsive cells within the kidney has implicitly caused sodium retention (even though the [Na^+] is not above the reference range) and potassium excretion. The cortisol at midnight should be much lower and the lack of a response to the dexamethasone suppression test suggests autonomous production of the hormone.

Compare the data for this patient with those in case 5 on page 178.

SECTION 2.xxvi

Disturbances of calcium or magnesium balance

Plasma calcium concentrations, or to be precise plasma ionised calcium concentrations, are closely monitored physiologically as described in Section 2.viii. Abnormalities in calcium homeostasis are not uncommon and need to be identified if potential long-term damage to tissues, especially kidneys and bone, is to be avoided. As is the case with abnormalities in Na^+ and K^+ balance, signs of hypercalcaemia or hypocalcaemia are not necessarily specific.

Hypercalcaemia

Clinical signs and symptoms of raised plasma calcium manifest in: the gastrointestinal tract (loss of appetite, even anorexia, nausea, vomiting, constipation); the kidney (polyuria with associated polydipsia, renal colic from stone formation), and the CNS (inability to concentrate). In addition to renal calculi, long-term effects may be evident on the heart, pancreas and skeleton (demineralisation causing bone pain).

By far the commonest causes of hypercalcaemia are hyperparathyroidism and malignancy. Primary hyperparathyroidism is characterised by autonomous over-production of PTH, usually from an adenoma in just one of the four parathyroid glands which may occur as part of a multiple endocrine neoplasia syndrome (MENS) involving either the pituitary and pancreas (MENS type 1), or thyroid, adrenal medulla and thyroid (MENS type 2a). Rarely are all four parathyroid glands hyperactive. Tertiary hyperparathyroidism develops in chronic conditions, typically chronic renal injury, when a patient has been

Essential Fluid, Electrolyte and pH Homeostasis, First Edition. Gillian Cockerill and Stephen Reed.

subject to long periods of appropriate PTH secretion (secondary hyperparathyroidism) in response to reduced plasma calcium concentration.

Hypercalcaemia of malignancy is common in multiple myeloma (a tumour of immunoglobulin producing B-type lymphocytes) and breast cancer, as in both cases there are lytic lesions within the bone as a result of demineralisation. Other tumours such as those of lung, ovary and oesophagus produce PTH related peptide (PTHrP) which engages PTH receptors and so mimics the action of the true hormone.

Rarer causes of hypercalcaemia include vitamin D intoxication, use of certain diuretics and granulomatous diseases such as sarcoidosis. A rare genetic condition called familial benign hypocalciuric hypercalcaemia (FBHH) is the fault of a failure of the renal calcium-sensing mechanisms, resulting in excessive calcium reclamation from the glomerular filtrate.

Case studies

(1) Male Age: 54 yrs Presenting features included nausea and vomiting, polyuria with frequency during the night. The patient's wife had noted periods of drowsiness in her husband.

	On admission	Ref.
Plasma		
Total Ca	3.10 mmol/L	2.25–2.55 mmol/L
PO_4	0.71 mmol/L	0.8–1.2 mmol/L
Albumin	39 g/L	35–50 g/L
PTH	15.6 pmol/L	1–6 pmol/L

Interpretation: Primary hyperparathyroidism. Normal physiological negative feedback should ensure that PTH concentration is low whenever calcium concentration is high. In this case, the feedback seems to be faulty and the PTH is inappropriate for the calcium. The actual value for PTH in this case is very high, but any value around or just above the upper limit of the reference range should be considered suspicious if the calcium is high-normal. The reported clinical signs are consistent with hypercalcaemia.

(2) Male Age: 39 yrs Long-standing kidney disease.

	On admission	Ref.
Plasma		
Total Ca	2.60 mmol/L	2.25–2.55 mmol/L
PO_4	1.35 mmol/L	0.8–1.2 mmol/L
Albumin	41 g/L	35–50 g/L
PTH	10.5 pmol/L	1–6 pmol/L

Interpretation: Hyperphosphataemia and hypocalcaemia are common in chronic renal disease, and so a PTH response would be expected. Tertiary hyperparathyroidism as in this case arises when there is parathyroid hyperplasia, which develops from pre-existing secondary hyperparathyroidism as in chronic renal failure.

(3) Male Age: 67 yrs Pancreatic cancer.

	On admission	Ref.
Plasma		
Total Ca	2.85 mmol/L	2.25–2.55 mmol/L
PO_4	1.10 mmol/L	0.8–1.2 mmol/L
Albumin	38 g/L	35–50 g/L

Interpretation: There are two possibilities in this case: hypercalcaemia of malignancy, or hyperparathyroidism as part of MENS. In the latter situation, the plasma phosphate might have been expected to be somewhat lower as PTH has a phosphaturic effect as well as a calcium-raising action in the nephron.

(4) Female Age: 59 yrs Confirmed breast cancer.

	On admission	Ref.
Plasma		
Total Ca	3.20 mmol/L	2.25–2.55 mmol/L
PO_4	0.95 mmol/L	0.8–1.2 mmol/L
Albumin	43 g/L	35–50 g/L
PTH	1.5 pmol/L	1–6 pmol/L

Interpretation: In contrast to case 3 above, this is a classic hypercalcaemia of malignancy. Some primary cancers produce bone metastases ('bony secondaries'), but given the site of the tumour there is probably secretion of PTHrP (see page 118). The true PTH is low due to feedback inhibition.

(5)	Female	Age: 76 yrs	Had been prescribed anti-inflammatories by her GP some time previously for bone and joint pain which was assumed to be due to osteoarthritis. Her only other clinical note was of hearing impairment.

The patient was admitted to hospital for minor surgery (unrelated to her bone symptoms) and the pre-op screen revealed an elevated alkaline phosphatase activity. There was no suspicion of liver disease and laboratory results strongly suggested that the alkaline phosphatase was of bone origin, despite the fact that the calcium, phosphate and magnesium were all normal. X-rays of the skull, pelvis and spine showed typical lesions in the bones. A diagnosis of Paget's disease was made.

A normal total calcium concentration is usual in this condition except in patients who have been immobile for periods of time when there may be a slight hypercalcaemia. Elderly patients such as this lady may be poorly nourished and/or have reduced exposure to sunlight, so vitamin D deficiency could be considered. However, in osteomalacia, the usual consequence of vitamin D deficiency in the elderly, plasma calcium and phosphate are often reduced. Osteoporosis usually shows no changes in calcium, phosphate or alkaline phosphatase results.

Hypocalcaemia

Clinical presentation of acute hypocalcaemia often includes peripheral neurological signs such as numbness or tingling sensations, and muscular spasms, twitching, seizures or tetany. In extreme cases there may be cardiac arrhythmias. Longer-term effects are apparent in changes to bone density. Hypocalcaemia may be seen as a consequence of alkalosis.

Assuming spurious causes such as hypoalbuminaemia (low total calcium due to reduced protein binding but ionised calcium is usually normal) and poorly collected samples (Section 2.xix, page 163) can be excluded, the main causes of hypocalcaemia are: hypoparathyroidism (especially in pre-term neonates and in adults following neck or thyroid surgery), vitamin D deficiency, acute pancreatitis and renal failure. In acute renal injury, the degree of hypocalcaemia may be significant and is probably due to phosphate retention, whereas in chronic renal impairment there may be only mild hypocalcaemia possibly because there is hypersecretion of PTH (appropriate secondary hyperparathyroidism) or reduced 1,25 vitamin D3 synthesis (25-OH vitamin D 1-α hydroxylase is located in the kidney).

Case studies

(6) Male Age: 18 yrs Known to have renal impairment.

	On admission	Ref.
Plasma		
Total Ca	2.16 mmol/L	2.25–2.55 mmol/L
PO_4	1.05 mmol/L	0.8–1.2 mmol/L
Albumin	29 g/L	35–50 g/L
PTH	4.3 pmol/L	1–6 pmol/L

Interpretation: The nephrotic syndrome results in the excessive loss of protein, mainly albumin, in the urine. Given that about half of plasma total calcium is bound to albumin, the hypocalcaemia is not unexpected (see page 163). Corrected calcium can be calculated or measurement of ionised calcium would indicate that there is no primary defect with calcium homeostasis, as suggested by the normal PTH.

(7) Female Age: 45 yrs A previous history of thyroid surgery.

	On admission	Ref.
Plasma		
Total Ca	2.06 mmol/L	2.25–2.55 mmol/L
PO_4	1.8 mmol/L	0.8–1.2 mmol/L
Albumin	41 g/L	35–50 g/L
PTH	1.3 pmol/L	1–6 pmol/L

Interpretation: Primary hypoparathyroidism due to surgical damage to the parathyroid(s).

(8) Female Age: 45 yrs Originally from Saudi Arabia but now living in the north of England. The patient presents to her GP with bone pain. Clinically, she appears well but on questioning, the GP suspects she has a poor diet.

	On admission	Ref.
Plasma		
Total Ca	2.20 mmol/L	2.25–2.55 mmol/L
PO_4	0.9 mmol/L	0.8–1.2 mmol/L
Mg	0.55 mmol/L	0.7–1.1 mmol/L
PTH	7.2 pmol/L	1–6 pmol/L
25hydroxy vitD3	15 nmol/L	>75 nmol/L

Interpretation: Vitamin D deficiency is not at all uncommon, especially in Moslem women who keep themselves covered and thus have little exposure to sunlight. The low plasma calcium has caused an appropriate rise in PTH, but there is reduced absorption of calcium in the gut. Plasma calcium concentration is being maintained by bone demineralisation.

Magnesium

Both hypermagnesaemia and hypomagnesaemia present with neuromuscular signs. There may also be nausea and vomiting (hyper-) or tetany and mental signs such as depression (hypo-). Isolated abnormalities of plasma magnesium are uncommon. Typical causes are listed in Table 2.9.

Table 2.9 Hyper- and hypomagnesaemia

Hypomagnesaemia	Hypermagnesaemia
Reduced intake, poor intestinal absorption or following gut surgery	Increased intake e.g. some antacids contain $MgSO_4$
Chronic renal failure	Acute renal failure
Hyperaldosteronism	Hypoadrenalism
Hyperparathyroidism	
Metabolic acidosis	
Drugs	
Alcoholism	

Given its importance as the second most abundant intracellular cation, it is perhaps surprising that relatively few plasma magnesium estimations are performed in routine laboratories.

Case studies

(9) Female Age: 30 yrs Post organ transplant surgery. Her treatment regime includes anti-inflammatory drugs.

The only notable result in the biochemistry screen was a slightly low magnesium concentration (0.55 mmol/L).

The most likely cause is cyclosporin toxicity. Measurement of immunosuppressive drugs such as cyclosporin and FK-506 (tacrolimus) is now common practice for many laboratories, so signs of toxicity should be easily detected.

(10) Tim F. was obsessively health conscious, often resorting to self-medication if he felt the need. On this occasion he consulted his GP for a check-up. He appeared thin, pale and dehydrated.

	On admission	Ref.
Plasma		
Na^+	139 mmol/L	135–145 mmol/L
K^+	3.0 mmol/L	3.5–4.5 mmol/L
HCO_3^-	32 mmol/L	23–28 mmol/L
Ca^{2+}	2.39 mmol/L	2.25–2.55 mmol/L
Mg^{2+}	2.1 mmol/L	0.7–1.1 mmol/L

Interpretation: It transpired that Tim frequently used Epsom salts ($MgSO_4$) as a laxative to ensure he was 'clean on the inside'. The results indicate that he had effectively 'overdosed'. He admitted to having diarrhoea which would account for the low [K^+].

SECTION 2.xxvii

Disorders of iron homeostasis

Iron deficiency is a worldwide problem with an estimated 20% of the global population affected, and women and children in developing countries being the most vulnerable. Overt deficiency develops when iron losses exceed intestinal absorption. Specific causes include poor diet, gastrointestinal damage (malignancy, surgery or coeliac disease) or dysfunction (gastritis, inadequate acid production in stomach) or parasitic infection. Predictably, given the distribution of iron within the body, a classic sign of iron deficiency is anaemia, so measurement of haemoglobin is indicated. However, anaemia may arise from a number of different causes, many of which are not related to iron deficiency but to a long-standing underlying pathology, so-called anaemia of chronic disease (ACD), thus making the laboratory diagnosis of iron deficiency problematic.

Serum ferritin concentration (reference range 20–300 μg/L) reflects overall body iron status, and plasma soluble transferrin receptor (sTfR, reference 3.0–8.0 mg/L) is a useful marker of the cellular need for iron and intensity of erythropoiesis. When cells have the need to import more iron, they express TfR on their surface, some of which is sloughed off and enters the plasma as sTfR. The 'gold standard' test for assessment of iron storage is analysis of bone marrow, but this is not a convenient procedure for the patient to undergo.

The use of plasma ferritin concentration and, to a lesser extent, sTfR for the diagnosis of iron deficiency is complicated if the patient has an inflammatory condition arising from any underlying chronic disease, such as an infection, or has had recent surgery. Such physical challenges to the body initiate an 'acute phase response' (APR) which

Essential Fluid, Electrolyte and pH Homeostasis, First Edition. Gillian Cockerill and Stephen Reed.

is part of the body's defence mechanism to injury. One aspect of the APR is the over-production of a number of hepatic proteins including ferritin. As a result, serum ferritin concentration in chronic disease will appear normal even if the patient is anaemic.

States of iron overload are much less commonly encountered than deficiency. Increased iron intake can lead to iron overload especially if there is simultaneous chronic alcohol misuse. Genetic idiopathic haemochromatosis is an HLA-A3 associated condition which occurs more frequently in men than women and leads to hepatic cirrhosis, cardiac muscle damage, pancreatic damage and other endocrinopathies.

SECTION 2.xxviii

Self-assessment exercise 2.4

Case studies

Comment on and interpret the data for each of the following case studies and identify the underlying mechanism for the changes seen, wherever possible.

(1) Male Age: 50 yrs Long-term use of analgesics for rheumatoid arthritis. Consulted his GP because he has experienced frequent urination and noted the very pale appearance of the urine.

	On admission	Ref.
Plasma		
Na^+	149 mmol/L	135–145 mmol/L
K^+	4.0 mmol/L	3.5–4.5 mmol/L
Osmolality	312 mmol/kg	285–295 mmol/kg
Urine		
Osmolality	180 mmol/kg	basal
Osmolality	210 mmol/kg	following 12 hr water deprivation
Osmolality	190 mmol/kg	following subcutaneous administration of vasopressin

Essential Fluid, Electrolyte and pH Homeostasis, First Edition. Gillian Cockerill and Stephen Reed.

(2) Female Age: 2 yrs Vomiting for 2 days. Clinically dehydrated, pale and hypotensive with rapid pulse.

	On admission	Ref.
Plasma		
Na^+	158 mmol/L	135–145 mmol/L
K^+	2.5 mmol/L	3.5–4.5 mmol/L
Cl^-	85 mmol/L	97–107 mmol/L
HCO_3^-	42 mmol/L	23–28 mmol/L
Urea	4.0 mmol/L	3.0–7.0 mmol/L

(3) Male Age: 25 yrs Recovered semi-conscious from the sea having fallen overboard from his boat whilst on a fishing trip.

	On admission	Ref.
Plasma		
Na^+	151 mmol/L	135–145 mmol/L
K^+	3.7 mmol/L	3.5–4.5 mmol/L
Urea	4.1 mmol/L	3.0–7.0 mmol/L
Creatinine	90 μmol/L	80–120 μmol/L
HCO_3^-	26 mmol/L	23–28 mmol/L
Cl^-	119 mmol/L	97–107 mmol/L

(4) Male Age: 25 yrs Head injury after being knocked off his bicycle. Initially there were no obvious problems but a day or two later his partner noticed a tendency to slurred speech, a degree of disorientation and irritability in the patient.

Analysis of a blood sample showed a low sodium concentration but all other data were normal. On questioning at the hospital, the cyclist admitted that his urine output had been a little lower than usual and urine itself was quite dark in colour.

(5) Female Age: 40 yrs A&E admission following road traffic accident. Blood loss at scene replaced with colloid infusion.

	On admission	Ref.
Plasma		
Na^+	123 mmol/L	135–145 mmol/L
K^+	7.4 mmol/L	3.5–4.5 mmol/L
Urea	49.3 mmol/L	3.0–7.0 mmol/L
Creatinine	1825 μmol/L	80–120 μmol/L
HCO_3^-	18 mmol/L	23–28 mmol/L
Glucose	5.6 mmol/L	3.5–11 mmol/L non-fasting
Urine		
Output	4 litres/24 hrs	

(6) Female Age: 80 yrs Shortness of breath.

	On admission	Ref.
Plasma		
Na^+	128 mmol/L	135–145 mmol/L
K^+	3.5 mmol/L	3.5–4.5 mmol/L
Urea	6.3 mmol/L	3.0–7.0 mmol/L
Creatinine	118 μmol/L	80–120 μmol/L

(7) Female Age: 61 yrs Being investigated for severe bone pain.

	On admission	Ref.
Plasma		
Na^+	129 mmol/L	135–145 mmol/L
K^+	3.6 mmol/L	3.5–4.5 mmol/L
Cl^-	95 mmol/L	97–107 mmol/L
HCO_3^-	22 mmol/L	23–28 mmol/L
Urea	8.1 mmol/L	3.0–7.0 mmol/L
Anion gap	18 mEq/L	8–12 mEq/L
Total protein	115 g/L	55–70 g/L
Albumin	28 g/L	35–50 g/L
Serum electrophoresis	Heavy band in γ region	

(8) Female Age: 50 yrs Presented to her GP with vague symptoms of feeling 'unwell', nausea and vomiting. On examination she was hypotensive and showed dull grey-brown pigmentation especially in the creases of the hands and around the face. The doctor, suspecting Addison's disease, took blood and arranged for a referral to the consultant endocrinologist at the local hospital.

	On admission	Ref.
Plasma		
Na^+	128 mmol/L	135–145 mmol/L
K^+	6.2 mmol/L	3.5–4.5 mmol/L
HCO_3^-	18 mmol/L	23–28 mmol/L
Cl^-	89 mmol/L	97–107 mmol/L
Urea	8.1 mmol/L	3.0–7.0 mmol/L
Creatinine	129 μmol/L	80–120 μmol/L
Osmolality	272 mmol/kg	285–295 mmol/kg
9am cortisol	100 nmol/L	>150 nmol/L

(9) Female Age: 35 yrs Recurrent back pain, described as often quite 'sharp'.

	On admission	Ref.
Plasma		
Total Ca	2.75 mmol/L	2.25–2.55 mmol/L
PO_4	0.85 mmol/L	0.8–1.2 mmol/L
Urine Ca	8.9 mmol/24 hrs	4.5–6 mmol/24 hrs

(10) Male Age: 25 yrs Undergoing 'routine' medical screening prior to taking up a new job. No presenting symptoms.

	On admission	One month later	Ref.
Plasma			
Total Ca	2.74 mmol/L	2.69 mmol/L	2.25–2.55 mmol/L
Urine Ca		0.5 mmol/L	1–3.5 mmol/L
PTH		4.6 pmol/L	1–6 pmol/L

(11) Patient known to have congestive cardiac failure on thiazides diuretics, but who by his own admission has not had a blood test 'for quite some time'.

	On admission	Ref.
Plasma		
Na^+	141 mmol/L	135–145 mmol/L
K^+	3.0mmol/L	3.5–4.5 mmol/L
HCO_3^-	24 mmol/L	23–28 mmol/L
Total Ca	2.54 mmol/L	2.25–2.55 mmol/L
Mg	0.5 mmol/L	0.7–1.2 mmol/L

SECTION 2.xxix

Summary of Part 2

- Homeostatic controls (often hormonal) ensure that in health the volumes of body fluids and their chemical compositions are maintained within very strict limits.
- Many minerals (anions and cations) are required for normal healthy cell function, often because of an association with the activity of key enzymes.
- Sodium, the principal cation in ECF, is concerned with osmoregulation: over- or under-hydration of cells could lead to serious consequences. Plasma sodium concentration also has a bearing on blood pressure.
- Potassium influences cell membrane potentials, and abnormalities lead to irregular neurological and muscular activities. Most of the total body potassium is within cells.
- The skeleton is rich in calcium, magnesium and phosphate. Intracellular calcium is involved with signalling, magnesium potentiates the action of insulin and is implicated with blood pressure regulation, and phosphate is an intracellular (and urinary) buffer.
- Hormones such as aldosterone, ADH, natriuretic factors, PTH, insulin and hydroxyvitamin D3 are key players in mediation of fluid and electrolyte homeostasis.
- The clinical signs and symptoms of hyper- and hyponatraemia can be difficult to differentiate so laboratory monitoring is required.
- Hypernatraemia is usually due to relative water deficit and hyponatraemia due to relative water excess.
- Hyponatraemia is commonly seen in patients especially soon after surgery.
- Potassium homeostasis is intimately linked with pH regulation.
- Hyperkalaemia may be caused by acute renal injury, hypoaldosteronism or redistribution between ICF and ECF.

Essential Fluid, Electrolyte and pH Homeostasis, First Edition. Gillian Cockerill and Stephen Reed.

- Hypokalaemia is most commonly as the result of diuretic therapy, but mineralocorticoid excess is a significant cause also.
- Hypercalcaemia is most frequently the result of hyperparathyroidism or malignancy.
- Hypocalcaemia is usually due to hypoparathyroidism (especially in pre-term neonates), renal failure, pancreatitis or vitamin D deficiency.
- Many disorders of electrolyte homeostasis are genetic, so family histories are informative.

Answers to Part 2 self-assessment exercises

Self-assessment exercise 2.1

1. The estimated glomerular filtration rate for a subject was 90 mL/min.
   (i) What volume of glomerular filtrate is produced in 24 hours?

$$24 \text{ hrs} \times 60 \text{ minutes} = 1440 \text{ minutes}$$

$$90 \times 1440 = 129{,}600 \text{ mL} = 129.6 \text{ litres}$$

   (ii) If cardiac output is 5 litre/min, what volume of whole blood arrives at the kidneys each minute?

   1 litre/minute, assuming 20% of the cardiac output goes to the kidneys. 1.25 litres if 25% of cardiac output reaches the kidneys.

   (iii) Use your answer to (ii) above to calculate the volume of plasma which arrives at the kidneys per minute, assuming that the volume taken up by red cells, white cells and platelets is 40% of the total blood volume.

$$1 \text{ litre} \times 0.6 = 600 \text{ mL/min}$$

   This answer can also be obtained thus:

   90 mL/min filtered ÷ 600 mL/min arriving at glomeruli = 15%

Essential Fluid, Electrolyte and pH Homeostasis, First Edition. Gillian Cockerill and Stephen Reed.

(iv) Use your answers to (i) and (iii) above to estimate the fraction of plasma that is completely filtered at the glomeruli per minute.

If 600 mL of plasma arrive at the kidneys each minute the total volume of plasma passing through the kidneys is:

$$600 \times 1440 = 864000 \text{ mL} = 864 \text{ litres}$$

Of this volume, 129.6 litres are filtered:

$$100 \times 129.6/864 = 15\%$$

(v) Assuming each kidney contains 1 million fully functional nephrons, what volume of glomerular filtrate passes through one nephron per day?

129.6 L glomerular filtrate are produced in 24 hours
= 129,600 mL ÷ 2,000,000 nephrons in both kidneys
= 6.45×10^{-2} mL = ~65 μL

(vi) Assuming mid-range plasma values for $[Na^+]$ and $[K^+]$, calculate the quantity of each mineral filtered per 24 hours (refer to Table 1.3 for typical values).

129.6 litres each containing 140 mmol Na^+ are filtered
129.6 × 140 = 18144 mmol = 18.1 moles Na^+ filtered

Each litre of glomerular filtrate contains 4.0 mmol K^+

129.6 × 4 = 518.4 mmol = ~0.52 moles K^+ filtered

(vii) Refer to Tables 1.3 and 1.6. Taking mid-range values for Na^+ and K^+ excretion in urine, and using your answer to (vi) above, estimate the quantity of each reabsorbed per 24 hours.

18.1 moles of Na^+ filtered in 24 hours
assuming 100 mmol (= 0.1 moles) Na^+ excreted in 24 hrs
∴ 18 moles Na^+ are reabsorbed.

0.52 moles of K^+ filtered in 24 hours
assuming 50 mmol (0.05 moles) K excreted in 24 hrs
∴ 0.47 moles K are reabsorbed.

2. Parathyroid hormone reduces the T_{max} value for phosphate. What effect would this have on phosphate excretion in the urine?

More phosphate would be excreted. T_{max} is the maximum reabsorption, so any reduction means more of the solute is lost. PTH has a phosphaturic effect.

3. Some subjects who have a genetic defect causing 21-hydroxylase deficiency in the adrenal cortex have a 'salt-losing' condition. Explain why this is the case.

   21-hydroxylase deficiency results in a lack of aldosterone and cortisol so Na is not reabsorbed in the nephron.

4. Why might a patient with *chronic* liver disease show signs of oedema?

   A reduction in the plasma protein, principally albumin, concentration would reduce the oncotic pressure so fluid would not be drawn back into the venules of the capillary system.

5. What would be the physiological effect of a lack of ADH (vasopressin)?

   The loss of large volumes of water as urine. Lack of ADH means that the aquaporins remain closed, meaning that water cannot be reabsorbed.

Were your answers correct? If not, try re-reading Sections 2.i to 2.iii.

Self-assessment exercise 2.2

1. Which of the following statements is/are true? (there may be more than one correct answer).
   (a) ADH and aldosterone exert their effects throughout the length of the nephron;
   (b) ADH acts mainly in the proximal region of the nephron *and* aldosterone is effective mainly in the distal part of the nephron;
   (c) ADH acts mainly in the distal part of the nephron *and* aldosterone is effective mainly in the proximal region of the nephron;
   (d) ADH and aldosterone both operate only in the proximal region of the nephron;
   (e) ADH and aldosterone both operate only in the distal region of the nephron.

   Only (e) is correct.

2. Aldosterone targets which of the following cell types?
   (a) α-intercalated cells
   (b) β-intercalated cells
   (c) principal cells
   (c) is correct
3. Name the two types of transport protein required to reabsorb sodium from the glomerular filtrate to the bloodstream.
   NKCC or ENaC (luminal face of tubular cell) and Na/K ATP'ase (basolateral face).
4. Which of the following statements can be correctly used to complete the sentence '*Most* sodium reabsorption in the nephron . . .'? (there may be more than one correct answer)
   (a) is independent of aldosterone;
   (b) occurs in the proximal tubules;
   (c) is carrier-mediated;
   (d) is influenced by the glomerular filtration rate, i.e. the volume of fluid passing through the nephron.
   All are correct.
5. A lack of aldosterone is likely to result in which of the following? (there may be more than one correct answer to this question)
   (a) an increase in the plasma concentrations of both Na *and* K;
   (b) a decrease in plasma [Na] and an increase in plasma [K];
   (c) a decrease in the plasma concentrations of both Na *and* K;
   (d) an increase in plasma [Na] *and* a decrease in urine[K];
   (e) an increase in urine [Na] *and* a decrease in urine [K].
   (b) and (e) are correct.
6. Rare renin-secreting tumours are a cause of hypertension. TRUE or FALSE. Explain your answer.
   True. Inappropriate production of renin leads to over-activity of the RAA (Figure 2.11). Increased Na reabsorption due to aldosterone stimulation leads to ADH secretion and thus an increase in blood volume, which causes hypertension.
7. Factors such as hypoxia (oxygen debt) or metabolic poisons that impair ATP generation will result in hyperkalaemia. Explain why this should be the case.
   The distribution of Na and K across cell membranes is dependent upon the Na/K ATP'ase pump. If ATP generation is compromised, the pump does not work efficiently, so K is not taken back into cells.

8. How can we explain mechanistically the link between a rise in blood pH and hypokalaemia?

   This association is usually explained by the movement of K into cells as protons move out of cells. Reciprocal movement of H^+ and K^+ is common.
9. Plasma concentrations of sodium, potassium, calcium, magnesium and iron all give a good indication of the total body content of each mineral. TRUE or FALSE. Explain your answer.

   False. Potassium, magnesium and iron are mainly sequestered within cells so their plasma concentrations do not reflect total body content. Plasma concentrations of sodium and calcium are good indicators of body load.
10. Which of the following is/are true of plasma iron? (there may be more than one correct answer to this question)
    (a) most Fe in the circulation is bound to ferroportin;
    (b) most Fe in the circulation is bound to transferrin;
    (c) most Fe in the circulation is bound to transferrin receptor;
    (d) most Fe in the circulation is bound to ferritin;
    (e) most Fe in the circulation is in the reduced form.

    (b) and (e) are correct.
11. Comment on the following set of data obtained on a plasma sample.

Total calcium	2.18 mmol/L	(ref: 2.25–2.55 mmol/L)
Phosphate	1.0 mmol/L	(ref: 0.8–1.2 mmol/L)
PTH	5.0 pmol/L	(ref: 1–6 pmol/L)
Albumin	31 g/L	(ref: 35–50 g/L)

The apparent hypocalcaemia is due to the low plasma albumin concentration. Applying the formula given in Section 2.vi shows that the corrected value is 2.38 mmol/L. The PTH value is not elevated because we assume the ionized calcium concentration is normal, so the Ca-sensing receptor response is not activated.

If you found the right answers, return to the text and continue. If not, you may need to read Sections 2.v to 2.xi again.

Self-assessment exercise 2.3

1. You are able to select from two possible ISEs for the estimation of Na in plasma. For one the Eisenmann-Nickolsky is quoted as

5.2×10^{-5} and for the other the same parameter has a value of 8.7×10^{-7}. Which of the two would you choose to use? Give reasons to support your answer.

The lower the value for the Eisenmann-Nickolsky constant the better the ISE, i.e. the more selective it is, and therefore less prone to interference by other ions. Therefore the second electrode is the better one (by approximately 2 orders of magnitude).

2. The reference range for pH of arterial whole blood is 7.35–7.45. What would be the difference in electromotive force (EMF) generated by the ISE at these two limits?

   The difference is 0.1 pH units. A 10-fold change in $[H^+]$ (one pH unit) would be equivalent to a difference of 59 mV, so 0.1 pH unit is only 5.9 mV. Very highly accurate and precise electrodes are therefore needed if reliable laboratory pH data are to be generated.
3. The measured osmolality of a plasma sample is 330 mmol/kg, yet the calculated osmolarity is 305 mmol/L. How can the difference be explained? (*Hint: review Section 1.iii*)

   The difference is greater than the expected osmol gap and is explained by the presence of an unmeasured substance which is likely to be osmotically active.

If you need to check any of your answers, return to Sections 2.xiii to 2.xvi.

Self-assessment exercise 2.4

1. Nephrogenic diabetes insipidus. Long-term analgesic treatment has caused ADH insensitivity. The urine is very dilute and neither fluid deprivation nor administration of ADH is able to have a significant impact on the water-retaining capacity of the kidneys.
2. Dehydration (high Na and high urea) is due to loss of hypotonic fluid. Low BP causes an aldosterone response hence low K. Loss of gastric Cl leads to an increase in bicarbonate to maintain total anion concentration.
3. The patient had presumably ingested a large volume of sea-water, in this case of near-drowning.
4. The blow to the head has caused over-secretion of ADH, so this is a dilutional hyponatraemia. Compare this with the case of hypernatraemia due to traumatic head injury.

5. These data and the clinical background suggest acute tubular necrosis due to hypovolaemia and hypotension following the blood loss before restorative treatment could be given.
6. This lady has congestive cardiac failure (CCF) with oedema. The reduced cardiac output causes stimulation of the renin-angiotensin-aldosterone system and consequently an increase in the ECF volume, so there is some evidence of ECF dilution. Diuretics are often used to treat such cases but care is needed not to cause hypokalaemia.
7. The abnormal protein results indicate that this must be a case of myeloma. The hyponatraemia is artefactual arising from volume displacement by the abnormal proteins. The calculated anion gap reflects the fact that proteins at normal blood pH carry a net negative charge, so the chloride and bicarbonate concentrations are reduced to ensure electrical neutrality.
8. A provisional diagnosis of Addison's disease was confirmed by these findings. Deficient secretion of adrenal mineralocorticoid (aldosterone, although cortisol has a mild effect) results in reduced sodium reabsorption coupled with reduced potassium secretion in the distal regions of the renal tubule. Loss of Na and water via the kidney results in hypovolaemia and stimulates an ADH response, further exacerbating the hyponatraemia.
9. Cause of the hypercalcaemia was not firmly established but most likely hyperparathyroidism or a genetic condition, although there is no family history in this case. The raised plasma and urine calcium concentrations are predisposing factors to the formation of renal calculi (stones), which is consistent with the clinical signs of lower back pain.
10. A tentative diagnosis of hypocalciuric hypercalcaemia was made. This is a benign condition but first-degree relatives should be investigated.
11. Drug-induced. Certain diuretics affect the renal handling of potassium and magnesium in particular but may have only a little impact on sodium.

Compare your answers with Sections 2.xviii to 2.xxvii and then continue with the text.

PART 3

Acid-base homeostasis

Overview

Part 3 of the text follows the same pattern Part 2. We begin by recapping basic chemistry of acidity and pH. Normal physiological mechanisms in which protons are generated during the daily acid challenge is first considered, and then how these ions are dealt with, mainly by the kidneys. The role of the respiratory system in regulating carbonic acid concentration via control of PCO_2 is discussed. There will be some recap of the basic buffering processes with particular emphasis on bicarbonate, proteins and phosphate. The concept of the 'strong ion difference' (SID), an alternative to the traditional bicarbonate-based approach to understanding metabolic acid-base upsets, will be introduced, explained and contrasted with interpretations based on application of the Henderson-Hasselbalch equation. The various disorders of acid-base homeostasis are described under the broad headings of respiratory and non-respiratory causes, and illustrated by case studies.

Essential Fluid, Electrolyte and pH Homeostasis, First Edition. Gillian Cockerill and Stephen Reed.

Normal physiological processes

SECTION 3.i

Acidity, pH and buffers: recap of some basic chemistry

Acidity is a measure of the proportion of protons and base anions present in a solution; an acidic solution has more protons than base anions, and vice versa for an alkali. Do not assume that an alkali is 'proton-free' as this is not true; at a pH above 7.0, the protons are merely outnumbered by base anions.

The relative strength of weak acids is determined by the extent of their dissociation, which is itself quantified by a K_a value; the higher the K_a, the stronger the acid. For convenience, the K_a is often expressed as pK_a:

$$pK_a = -\log_{10}[K_a]$$

The pK_a value defines the pH at which the acid group is 50% dissociated. At pH values below pK_a, the weak acid is predominantly protonated, whilst if the local pH is greater than the pK_a value, the acid group is mostly in the de-protonated form. The actual pK_a value gives a measure of the tendency of the weak acid to donate its proton; the lower the value, the stronger the acid (i.e. greater tendency to dissociate) and vice versa. Strong mineral acids such as HCl or H_2SO_4 are fully ionised and therefore have very low pK_a values.

Water is a very weak acid, that is, the dissociation equilibrium lies towards the left, so, for water:

$$H_2O \rightleftarrows H^+ + OH^-$$

Essential Fluid, Electrolyte and pH Homeostasis, First Edition. Gillian Cockerill and Stephen Reed.

The extent of ionisation is given by the symbol K_w and experiment has shown a value, at 25°C, for the dissociation constant to be 10^{-14} (i.e. $[H^+] \times [OH^-] = 10^{-7} \times 10^{-7}$).

$$pK_w = -\log[K_w] = 14$$
$$\text{and} \quad pH + pOH = 14$$

Given: $pH = -\log[H^+]$ or, $\log \frac{1}{[H^+]}$

where $[H^+] = H^+$ concentration in *mol/L*

$$\text{so,} \quad \text{at pH 7 } [H^+] = [OH^-]$$
$$\text{and} \quad \text{at pH 7 } [H^+] = 1 \times 10^{-7} \text{ mol/L} = 100 \text{ nmol/L}$$

Note that at 37°C (body temperature), $pK_w \sim 13.7$, indicating that neutral pH ~ 6.85, *not* 7.00, a value that correlates well with experimental evidence that ICF pH ~ 6.9. It is the pH of ICF that is ultimately the most important physiologically, since most biochemical reactions occur within cells rather than in the circulation. Measurement of arterial blood pH is a convenient surrogate for assessing ICF pH.

Each successive proton dissociation of a weak acid will have a particular pK_a. A polyprotic acid such as phosphate H_3PO_4 can donate, one at a time, 3 protons, so has 3 pK_a values. The pH value, because it is measured on a logarithmic scale with no units, may disguise the real magnitude of the change in hydrogen ion concentration. For example, a decrease in pH of only 0.3 units represents a *doubling* of the proton concentration because log 2 = 0.30; conversely, an increase in pH of 0.3 units is equivalent to a halving of the $[H^+]$ because log 1/2 = −0.3.

Two useful 'rules of thumb' approximations

(1) For every 0.01 pH unit change from 7.00, the $[H^+]$ changes by 1 nmol/L. This 'rule' applies most reliably in the range of pH values 7.20 to 7.60.

(2) For every 0.1 unit increase in pH from 7.00, multiply the previous $[H^+]$ by 0.8 – 'the 80% (or 0.8) approximation'. For example:

$$\text{pH } 7.00 = 100 \text{ nmol/L}$$
$$\text{pH } 7.10 = 100 \times 0.8 = 80 \text{ nmol/L}$$
$$\text{pH } 7.20 = 80 \times 0.8 = 64 \text{ nmol/L}$$

Buffers

The pH of the body fluid compartments is very carefully regulated, as changes can have an impact on the normal metabolic activities of a cell. Without appropriate homeostatic buffering mechanisms, the protons produced by metabolism would make survival impossible. Physiologically important buffers consist of a conjugate weak acid and its corresponding base anion. The relative concentrations of the components of a buffer system at a given pH are defined by the Henderson-Hasselbalch equation:

$$pH = pK_a + \log_{10}\left\{\frac{[\text{base}]}{[\text{conj. weak acid}]}\right\}$$

We can predict from the Henderson-Hasselbalch equation that buffers are most efficient in 'mopping-up' free protons when there are approximately *equal* proportions of base and acid present, i.e. 1:1 acid: base ratio and $pH = pK_a$. In practice, in a test-tube, buffer efficiency is maximal when the actual $pH = pK_a \pm 1$ pH unit. Some *in vivo* buffer systems, notably the carbonic acid/bicarbonate couple, work effectively outside the range of $pK_a \pm 1$ pH unit because we are able, through normal physiological mechanisms, to maintain the correct base:acid ratio.

As we shall see in more detail later, the principal buffers in body fluids are proteins and bicarbonate/carbonic acid in blood, bicarbonate/ carbonic acid in ISF, proteins and phosphate in ICF, and phosphate and ammonia in urine.

SECTION 3.ii

Some worked example calculations

Given that $\text{pH} = -\log[\text{H}^+]$ or $\log \dfrac{1}{[\text{H}^+]}$

and $\text{pH} = \text{pK}_a + \log_{10}\left\{\dfrac{[\text{base}]}{[\text{conj. weak acid}]}\right\}$

(i) A sample of urine has a hydrogen ion concentration of 85 μmol/L. What is the pH?

$$85\ \mu\text{mol/L} = 85 \times 10^{-6}\ \text{mol/L} = 8.5 \times 10^{-5}\ \text{mol/L}$$

$$\text{pH} = \log 8.5 \times 10^{-5} = -4.07$$

$$\text{so,}\quad \text{pH} = 4.07(\sim 4.1)$$

This is a fairly low value as urine typically has a pH in the range 5.5 to 6.5 depending upon diet.

(ii) In health, arterial blood has a pH of 7.40 (± 0.05). What is the $[\text{H}^+]$?

If pH is 7.40, then we can say, $[\text{H}^+] = 10^{-7.4}$ mol/L (note the unit)

Mathematically, a simple way to perform the calculation is:

$10^{-7.4} = 10^{-8} + 10^{+0.6}$

take antilog $10^{+0.6} = 3.98$, and retain the 10^{-8}

Essential Fluid, Electrolyte and pH Homeostasis, First Edition. Gillian Cockerill and Stephen Reed.

(*Note*: on most calculators, the antilog is found using 'inverse log' or '2nd function log')

so, $[H^+] = 3.98 \times 10^{-8}$ mol/L

or, $[H^+] = 39.8 \times 10^{-9}$ mol/L = 39.8 nmol/L

~ 40 nmol/L

Alternatively, and even easier in terms of using a calculator:

$[H^+]$ nmol/l = antilog 10^{9-pH} (note the unit, nmol/L)

antilog (9 – 7.40)

antilog 1.6

$\therefore [H^+] = 39.8$ nmol/L ~ 40 nmol/L

Compare the value of 40 nmol/L with the concentrations of sodium, potassium, calcium etc. given in Table 1.3 (Section 1.viii), taking careful note of the units for each value. The reference range for arterial blood pH is 7.35–7.45 which corresponds to 35–45 nmol/L $[H^+]$. Implicitly, therefore, physiological control of pH is extremely sensitive, operating as it does at concentrations which are much lower than those of, for example, glucose.

(iii) Assuming that the only buffer in urine is phosphate ($H_2PO_4^-$/HPO_4^{2-}) and given the pK_a of dihydrogen phosphate is 6.8, what is the pH of urine if the $[H_2PO_4^-]$ is 2.5 times higher than the $[HPO_4^{2-}]$?

$$pH = pK_a + \log\left\{\frac{[\text{base}]}{[\text{conj. weak acid}]}\right\}$$

$$pH = 6.8 + \log(1/2.5)$$

$$pH = 6.8 + \log 0.4$$

$$pH = 6.8 + (-0.3979)$$

$$\therefore pH = 6.4$$

As a quick check, recall that if the base:acid ratio had been 1 to 2 rather than the stated ratio of 1 to 2.5, the pH would have been 0.3 pH units lower than the pKa. The calculated value above is therefore plausible.

Remember also that if the [base] > [acid] the pH must be higher (more alkaline) than the pK_a. In this example, [base] < [acid] so the final pH is lower than the pK_a. Furthermore, we can predict that if the base:acid ratio had been 2:1, the pH would have been $6.8 + 0.3 = 7.1$, and if the base:acid ratio were 2.5:1 the final result would have been $6.8 + 0.3979 \sim 7.2$

(iv) The pK_a of carbonic acid is 6.10. Calculate the base:acid ratio in arterial blood at normal pH.

$$H_2CO_3 \rightleftarrows H^+ + HCO_3^-$$

Conjugate weak acid ↗ (H₂CO₃); base anion ↖ (HCO_3^-)

Substituting into the Henderson-Hasselbalch equation;

$$7.40 = 6.1 + \log [HCO_3^-]/[H_2CO_3]$$
$$7.40 - 6.1 = \log [HCO_3^-]/[H_2CO_3]$$
$$\therefore 1.30 = \log [HCO_3^-]/[H_2CO_3]$$
$$\text{antilog } 1.3 = [HCO_3^-]/[H_2CO_3]$$
$$19.95 = [HCO_3^-]/[H_2CO_3]$$

Therefore, at a normal blood pH, the $[HCO_3^-]/[H_2CO_3]$ is ~20:1, which is clearly very different from the 'ideal' ratio of 1:1 for optimal buffering effect. In a test-tube, this would not be an efficient buffer system. but because both $[HCO_3^-]$ and $[H_2CO_3]$ can be regulated physiologically, it is possible to maintain the apparently unfavourable ratio.

Details of the mechanisms that regulate $[HCO_3^-]$ and $[H_2CO_3]$ are described later in Section 3.vi.

SECTION 3.iii

Self-assessment exercise 3.1

1. (a) Confirm that the reference range for arterial blood $[H^+]$ is 35–45 nmol/L.
   (b) What is the $[H^+]$ gradient across the plasma membranes of cells in contact with interstitial fluid (assume this to be identical in composition to venous blood)?
2. Verify that $pH = pK_a$ when [base] = [conjugate acid].
3. Given pK_a for lactic acid = 3.8, what is the lactate:lactic acid ratio inside a cell under normal biochemical conditions?
4. The normal range for arterial blood pH is 7.35 – 7.45 but individuals are able to survive pH values as low as 7.10 or as high as 7.70.
   (a) What is the $[H^+]$ in a blood sample whose pH is 7.10?
   (b) What is the $[H^+]$ in a blood sample whose pH is 7.70?
5. Given that pK_a for H_2CO_3 *in vivo* = 6.10, calculate the base to acid *ratio* at:
   (a) the upper limit of normal for arterial pH.
   (b) the lower limit of normal for arterial pH.
6. An astute junior doctor notices that the $[H^+]$ of an arterial blood sample is reported as 65 nmol/L but the pH is given as 7.25. Is she justified in asking for the result to be confirmed?
7. Urine contains approximately the same amounts of phosphate and sulphate, yet only phosphate is able to act as an effective buffer. Suggest a reason for this.

Essential Fluid, Electrolyte and pH Homeostasis, First Edition. Gillian Cockerill and Stephen Reed.

SECTION 3.iv

Homeostasis and the 'daily acid challenge'

As described in Part 2 of this text, control of sodium and potassium concentrations in plasma is rigorous in order to maintain normal ECF values within the ranges of 135–145 mmol/L and 3.5–4.5 mmol/L respectively. Significant deviation from these ranges can have serious consequences for cell function and therefore health. Although we recognise that significant localised pH differences exist within cells, for example across the mitochondrial membrane (according to Mitchell's chemi-osmotic hypothesis of ATP generation), a typical intracellular pH is 6.85 and any abnormalities in local ICF pH may have a great impact on the ability of enzymes to fulfil their normal function in the maintenance of metabolism. Interestingly, normal blood pH is more alkaline than that in the ICF at 7.40 (arterial) and 7.35 (venous). These values for ICF and ECF pH are representative of the 'pool' of protons that exists in body fluids. The size of this pool is actually quite small, but the turnover of protons within the pool on a daily basis is, in comparison, very large, so appropriate physiological mechanisms need to be in place to prevent extreme changes in pH.

In contrast to ECF [Na^+] and [K^+], blood proton concentration is measured in the nmol/L range, approximately 3,500,000 times lower than normal [Na^+] and 100,000 times lower than normal potassium concentrations. Furthermore, the reference range for [H^+] is quite narrow ($\pm$ 5 nmol/L), so we can see that homeostatic regulation of pH needs to be very sensitive. Also, the potential for change in ECF pH brought about by normal metabolic processes is significant, especially when seen in comparison with possible changes in Na^+ and K^+ concentrations. Any acid challenge must be matched by the capacity

Essential Fluid, Electrolyte and pH Homeostasis, First Edition. Gillian Cockerill and Stephen Reed.

of the homeostatic response, and indeed even relatively large changes (e.g. by a factor of 2) in ECF [H^+], although serious, are physiologically tolerable, whereas comparable relative changes in plasma [Na^+] are incompatible with life. The range of pH values which may reasonably be encountered in practice is 7.00 to 7.80; this represents a 6.5-fold change from 15 to 100 mmol/L. So is not only pH regulation very sensitive, it also has a much wider dynamic range than any of the other electrolytes.

Essential metabolic processes generate large quantities of protons each day, which if left unchecked would jeopardise the very biochemical processes that produced them, and therefore the health of cells. The daily generation of protons from metabolism is very significant, but the net production is much lower as often protons are recycled within the metabolic 'pool'. The net quantity of acid added to the body fluids each day is in excess of 15 mol (not mmol) of protons, most of which comes from respiratory carbon dioxide production:

$$CO_2 + H_2O \longrightarrow H_2CO_3 \longrightarrow HCO_3^- + H^+ \qquad (1)$$

(this is a 'key' equation which we shall meet again)

Carbon dioxide is often referred to as a 'volatile acid' because it can be expired as a gas.

A small proportion (approximately 75 mmol/L) of the overall acid generated each day comes from metabolism of carbohydrates, fats and amino acids in the form of organic acids (e.g. lactate and acetoacetate, etc.). It is estimated that in excess of 1 mole of lactate is produced each day through normal metabolism, but the vast majority is itself metabolised mostly in the liver via the Cori cycle, in which lactate produced by muscle is transported to the liver to be used for the production of glucose via gluconeogenesis. The slight reduction in cellular pH that accompanies the lactate load inhibits glycolysis and simultaneously favours gluconeogenesis, which is, significantly, a process that actually generates a bicarbonate ion as oxaloacetate is converted to phosphoenolpyruvate for every lactate metabolised. The tricarboxylic acid (TCA) contains two oxidative decarboxylation steps, so these carbon dioxide molecules may also be hydrated to form bicarbonate. These processes are outlined in Figure 3.1.

We also generate inorganic acids such as sulphuric (from the metabolism of sulphur-containing amino acids) and phosphoric acids (from hydrolysis of phosphoesters). To draw a contrast between these

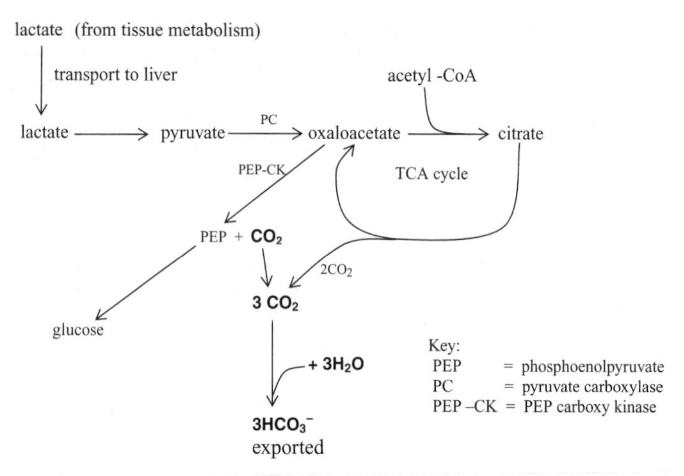

Figure 3.1 Lactate turnover in the liver generates bicarbonate

and carbon dioxide, such metabolic acids are called 'fixed' (or non-volatile) acids and are excreted in the urine.

The small figure of net metabolic acid production disguises the staggering turnover (liberation followed by consumption) of protons that occurs during normal metabolism. An adult consumes and re-generates nearly his own body weight of ATP each day! Each time an ATP molecule is hydrolysed or when an oxidation reaction involving NAD^+ occurs, a proton is liberated . . .

$$ATP^{4-} \longrightarrow ADP^{3-} + \text{phosphate} + H^+$$
$$NAD^+ + RH_2 \longrightarrow NADH + H^+ + R$$

. . . but most of these protons are consumed when ATP is resynthesised or when NADH is reoxidised. Any protons that are not re-used in metabolism need to be dealt with in some way to avoid excessive and probably lethal acidification of body fluids. It is these non-recycled protons that represent the net daily acid challenge.

The 'acid insult' that the body experiences every day is dealt with in a number of ways. For example, the acid may be diluted in body water: this clearly has limited efficiency, since very soon the pH of those fluids would fall dangerously low. Some of the protons can be exchanged for sodium or even calcium ions within the skeleton, but this mechanism alone would result in demineralisation and thus weakening of the bones. Most of the metabolic acid is initially buffered and then excreted. Loss of carbon dioxide effectively converts some free protons into water by reversal of Equation 1 shown earlier in this Section (page 229). Non-volatile acidic compounds (75 mmol/day) are excreted through the kidneys and the gut.

In addition to the quantity of protons added to the body fluids each day, there is a relatively small loss of base via the gut. Intestinal and in particular pancreatic fluids are alkaline, so faeces contain some bicarbonate ion. Only in cases of extreme diarrhoea is the loss of bicarbonate sufficient to affect overall acid-base homeostasis.

Summarising the key points, pH regulation is required at very low concentrations but needs to operate over a wide physiological range in order to cope with large additions of acid to the 'proton pool' within body fluids.

SECTION 3.v

Physiological buffering

In order to prevent cellular dysfunction, the acid load must be eliminated or neutralised. Several buffer systems can be identified in body fluids and classified generically as carbonate (mainly bicarbonate, but also small contributions from carbonate and carbamino compounds) and non-carbonate buffers. Specifically, these buffers are:

(i) BICARBONATE-CARBONIC ACID buffer in the plasma (ECF);
(ii) PROTEIN (mainly albumin) in the plasma (ECF) and intracellularly (ICF);
(iii) HAEMOGLOBIN (Hb) in the red blood cells (ICF); and
(iv) PHOSPHATE (Pi, in ECF, ICF but especially in urine).

i.e. (i) $\underset{\text{small buffer effect}}{CO_3^- + H^+} \longrightarrow \underset{\text{major buffer effect}}{HCO_3^- + H^+} \longrightarrow H_2CO_3$

(ii) $\text{proteinate}^{12-} + H^+ \longrightarrow \text{proteinate}^{11-}$

(iii) $H^+ + Hb \longrightarrow HHb$

(iv) $HPO_4^{2-} + H^+ \longrightarrow H_2PO_4^-$

Free H^+ is strong acid, but HHb, proteinate and H_2CO_3 are weak acids.

Bicarbonate and protein, including haemoglobin in whole blood, and phosphate are the major contributors to what is called the 'buffer base' in blood, and abnormal acid-base status should only be fully interpreted in terms of changes in total buffer base and not bicarbonate alone. The plasma concentration of phosphate is only about 1 mmol/L, thus its buffer capacity is limited. Phosphate is physiologically more important as a buffer of urine.

Protein buffering is especially important inside cells where the metabolic acids are being produced. Haemoglobin is a particularly efficient buffer because (a) it is present in red cells at a very high

Essential Fluid, Electrolyte and pH Homeostasis, First Edition. Gillian Cockerill and Stephen Reed.

concentration (150 g/L), and (b) because it is relatively rich in residues of the amino acid histidine which have pK_a values around 6.0 depending upon their position within the peptide chain, clearly within the physiological range of cytosolic pH. Similarly, because of their greater buffering capacity relative to bicarbonate, it is probable that plasma proteins are the first line of defence in the face of an acid insult, and bicarbonate replenishes the protein.

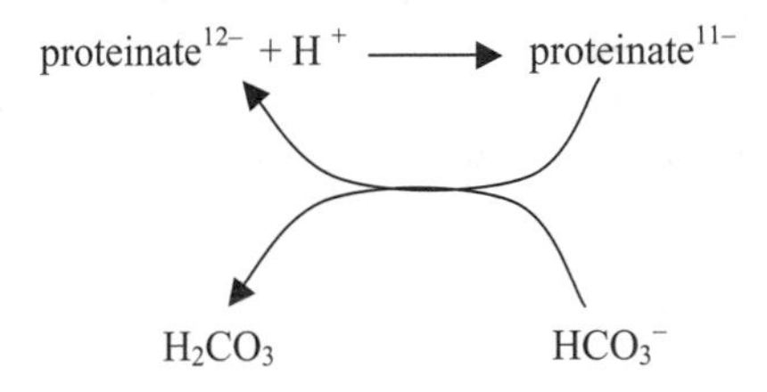

For each mmol/L of H^+ added to the plasma, 1 mmol/L of HCO_3^- is titrated (i.e 'removed' from the plasma). Overall, the bicarbonate system contributes more than 60% of the total buffering effect in blood, but very little within cells where proteins and phosphate are more important.

The carbonic acid/bicarbonate (H_2CO_3 / HCO_3^-) buffer system

Traditionally, the carbonic acid/bicarbonate buffer system is seen to be the axle around which acid-base balance revolves, and our understanding of physiological buffering of the ECF is based upon application of the Henderson-Hasselbalch equation. As stated above in Section 3.i, in a test-tube, chemical buffers are most effective when used for situations where the expected pH is ±1 pH unit from the pKa. The pK_a for H_2CO_3 *in vivo* is 6.1, which is more than 1 pH unit away from arterial blood pH (7.40). In a test-tube, the H_2CO_3/HCO_3^- buffer system would not work well at pH 7.40. Fortunately, we are not test-tubes and H_2CO_3/HCO_3^- is an extremely important buffer system in the ECF because we are able to regulate the concentrations of carbonic acid and bicarbonate.

The bicarbonate-carbonic acid buffer (see page 229) is efficient *in vivo* because the concentrations of base and acid are easily regulated by physiological processes. The CO_2 content of the blood is controlled by the rate and depth of breathing. Breathe quickly (hyperventilate) and you will *reduce* the CO_2 in your blood and pH rises. Hold your breath and your blood CO_2 will *increase*, resulting in a pH decrease. Moreover, the CO_2 content controls the rate and depth of breathing. An increase in blood CO_2 will cause an increase in respiration rate and a decrease in CO_2 will slow respiration. The relationship between gas content of the blood and respiration will be discussed more fully in Sections 3.vii and 3.viii. The concentrations of protons and bicarbonate are controlled largely by the kidneys, details of which will be provided in Section 3.vi.

Mathematically, the relationship between concentrations of the acid component and the base component of any buffer system is described by the Henderson-Hasselbalch equation:

$$pH = pK_a + \log\left\{\frac{[\text{base}]}{[\text{conj. weak acid}]}\right\}$$

which equates to:

$$pH = pK_a + \log\left\{\frac{[HCO_3^-]}{[H_2CO_3]}\right\}$$

Thus, physiologically, the relative concentrations of the acid component (H_2CO_3) and the base component (HCO_3^-) are regulated by different organs. The bicarbonate component is regulated by the kidney, whereas $[H_2CO_3]$ is related to the carbon dioxide content of the blood, the PCO_2 (partial pressure of carbon dioxide), and is thus controlled by the respiratory system and lungs.

Hence a fall in blood pH is due to either a rise in carbonic acid concentration or a fall in plasma bicarbonate concentration, whilst a rise in blood pH is the result of a decrease in carbonic acid concentration or a rise in plasma bicarbonate. Expressed another way, changes in blood pH are directly proportional to [bicarbonate] and inversely related to [carbonic acid]. Plasma concentration of HCO_3^- can be measured easily in the clinical laboratory or at the patient's bedside, but the $[H_2CO_3]$ concentration cannot be so easily measured. An indirect measure of carbonic acid concentration can be obtained, however, because $[H_2CO_3]$ is related to the solubility of CO_2 in water, i.e. $[H_2CO_3] \propto PCO_2$. Because CO_2 content is measured as partial pressure, the correct SI unit is kilopascal (kPa), although use of mmHg (millimetres of mercury) is still to be found.

The proportionality sign can be replaced by the value for the Bunsen solubility of carbon dioxide in water, 0.225 per kPa:

$$\text{e.g. if } PCO_2 = 5.3 \text{ kPa, then } [H_2CO_3] = 5.3 \times 0.225$$
$$= 1.2 \text{ mmol/L}$$

We can now re-write the Henderson Hasselbalch equation as:

$$pH = pK_a + \log_{10}\left\{\frac{[HCO_3^-]}{PCO_2 \times 0.225}\right\}$$

This is a *very* important equation to remember. Simplifying this equation shows the important relationships:

$pH \propto [HCO_3^-]$
$pH \propto 1/[H_2CO_3]$ which is the same as $pH \propto 1/[PCO_2]$.

Given the prime importance of the renal and respiratory systems in controlling the concentrations of the two components of the bicarbonate buffer, we can write:

$$pH \propto \text{kidney/lung}$$

as a useful *aide memoir* for interpreting acid-base pathology using the traditional model.

Worked examples

1. If $[HCO_3^-] = 20$ mmol/L and the $PCO_2 = 6.5$ kPa, we can calculate the blood pH:

$$pH = 6.1 + \log\left\{\frac{20}{6.5 \times 0.225}\right\}$$
$$= 6.1 + \log\left\{\frac{20}{1.46}\right\}$$
$$= 6.1 + \log 13.7$$
$$= 6.1 + 1.13$$
$$= 7.23$$

2. Calculate the blood pH if: $[HCO_3^-]$ = 18 mmol/L, PCO_2 = 4.0 kPa

$$\begin{aligned} pH &= 6.1 + \log\{18/0.225 \times 4.0] \\ &= 6.1 + \log\{18/0.89\} \\ &= 6.1 + \log 20.2 \\ &= 6.1 + 1.30 \;\therefore\; pH = 7.40 \end{aligned}$$

NB: the blood pH is normal although both $[HCO_3^-]$ and PCO_2 are reduced.

3. Calculate the blood pH if: $[HCO_3^-]$ = 30 mmol/L, PCO_2 = 4.5 kPa

$$\begin{aligned} pH &= 6.1 + \log\{30/0.225 \times 4.5] \\ &= 6.1 + \log\{30/1.01\} \\ &= 6.1 + \log 29.63 \\ &= 6.1 + 1.47 \;\therefore\; pH = 7.57 \end{aligned}$$

4. Calculate the $[H^+]$ if pH of pancreatic juice = 8.5

$$\begin{aligned} pH\ 8.5 &= 10^{-8.5}\,[H^+] \\ &= 10^{-9} + 10^{+0.5} \\ &= 3.16 \times 10^{-9}\ mol/L \end{aligned}$$

$\therefore [H^+] \sim 3.2$ nmol/L, compare with 40 nmol/L for whole blood.

SECTION 3.vi

The role of the kidney in acid-base homeostasis

An overview of renal function with a focus on electrolyte homeostasis was given in Part 2 (Sections 2.ii, 2.iii). Traditional models used to explain mechanisms of acid-base homeostasis place the bicarbonate-to-carbonic acid ratio at the centre of physiological control of blood pH, as maintenance of the critical 20:1 HCO_3^- :H_2CO_3 ratio which we calculated in Section 3.ii is seen to be pivotal in regulating blood pH. As shown in Figure 3.2 below, the kidney plays an important role in the regulation of the pH of body fluids via its ability to (a) reabsorb filtered HCO_3^-, (b) 'generate' HCO_3^- and (c) secrete protons into the glomerular filtrate. Bicarbonate passes into the renal vein from all parts of the nephron; acidification of the glomerular filtrate, however, is restricted to the distal regions of the nephron. Net renal excretion of acid amounts to approximately 70–75 mEq (equivalent to 70–75 mmol H^+) per day, which is equal to the amount of weak metabolic acid generated in the same time period to replace that lost to buffering the ECF.

Just take a moment at this point to compare the net amount of metabolic acid produced per day with the plasma [H^+]: 75 mmol/day compared with 40 nmol/L. If 75 mmol of protons were to enter the plasma (volume 3 litres, giving [H^+] of 25 mmol/L) each day without being buffered or excreted, the plasma pH would fall to less than 2; this indicates very clearly the extent of buffering capacity needed to deal only with daily metabolic acid. Furthermore, there is another source of protons yet to be considered in detail: carbonic acid (H_2CO_3) arising from carbon dioxide itself derived from cellular respiration.

Essential Fluid, Electrolyte and pH Homeostasis, First Edition. Gillian Cockerill and Stephen Reed.

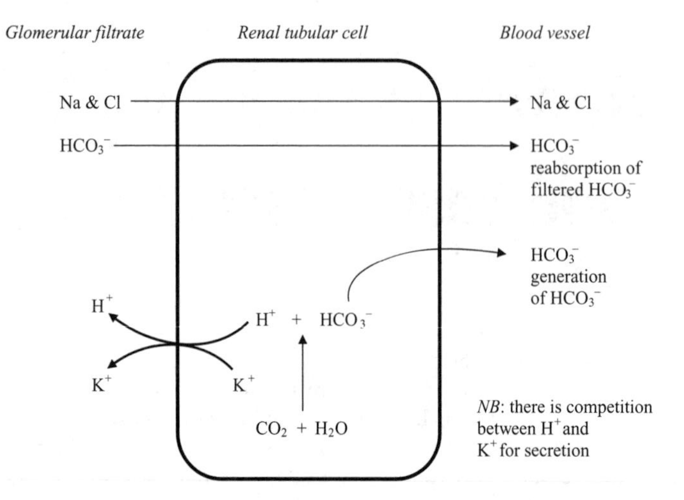

Figure 3.2 Simplified diagram to show the role of the nephron in acid-base balance

Even a brief consideration of Figure 3.2 shows the close inter-relationship between acid-base balance and electrolyte balance. Note the importance of: (a) sodium in allowing proton secretion and (b) the competition for secretion between K^+ and H^+, thus changes in the blood concentration of either one will affect the concentration of the other. Dealing with these two factors in turn:

(a) Urine is electrically neutral. In situations where there is the need to excrete an unusually high load of anion, for example ketoacids (chemically more correctly referred to nowadays as oxoacids), a counter-cation must also be excreted and the most likely candidate for this role is sodium. Thus, the origin of many cases of acidosis is as much the result of impaired proton secretion due to increased Na^+ and associated anion loss as it is to overproduction of metabolic acid.

(b) Changes in potassium concentration and proton concentration are inter-related:

Hyper<u>kal</u>aemia ($\Uparrow$ plasma $[K^+]$) $\longleftrightarrow$ hyper<u>*proton*</u>aemia (=low blood pH)
Hypo<u>kal</u>aemia ($\Downarrow$ plasma $[K^+]$) $\longleftrightarrow$ hypo<u>*proton*</u>aemia (=high blood pH)

Bicarbonate transport from glomerular filtrate to plasma is influenced by the electrochemical gradient across the membranes of the tubular cells. If the concentration of chloride (quantitatively the principal anion in plasma) rises or falls, so too will the electrochemical gradient, allowing more or less bicarbonate to be returned to the plasma. Thus, when plasma [chloride] is high, proportionately less bicarbonate is reabsorbed resulting in acidosis. Thus, renal handling mechanisms of Na^+, K^+, Cl^- and H^+ are intimately linked.

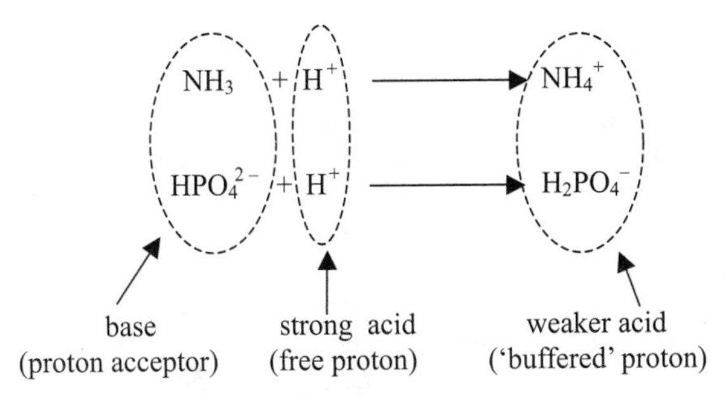

Figure 3.3 Ammonia and phosphate are bases

The net effect of proton secretion is to generate urine which normally has a pH of around 6.0–6.5, the acidity being due to the presence of, not free protons, but ammonium (NH_4^+) and so-called 'titratable acidity', which is dihydrogen phosphate ($H_2PO_4^-$) Figure 3.3.

The HPO_4^{2-} present in the luminal fluid of the proximal tubule is that which has been filtered from the blood plasma at the glomeruli. At typical plasma pH, the ratio of HPO_4^{2-} to $H_2PO_4^-$ is approximately 4:1, a figure that indicates there is substantial buffering capacity at least within the early part of the nephron. As the phosphate in the glomerular filtrate progresses along the nephron and protons are secreted, HPO_4^{2-} is converted to $H_2PO_4^-$. Acidification of the glomerular filtrate by secretion of NH_4^+ and $H_2PO_4^-$ requires the combined action of enzymes and transport proteins in all parts of the nephron.

Carbonic anhydrase in the renal tubule

Carbonic anhydrase is a zinc-containing enzyme found widely distributed in body tissues. This enzyme has a major role to play in acid-base homeostasis in the proximal and distal portions of the renal nephron and red blood cells. The action of carbonic anhydrase (CA, also called carbonate dehydratase) is to accelerate the hydration of CO_2:

$$CO_2 + H_2O \xrightarrow{CA} H_2CO_3$$

The carbonic acid so produced spontaneously dissociates to form H^+ and HCO_3^-.

Carbonic anhydrase occurs in two different forms (CA II and CA IV), both being found within the proximal tubule but in different specific locations which are (i) within the cytosol of tubular cells and (ii) attached to the luminal surface of the same cells. The combined actions of CA II and CA IV are required to bring about reabsorption of filtered bicarbonate, although HCO_3^- ions that are returned to the bloodstream are not the same ones that crossed the glomerular barrier.

Figure 3.5 shows how bicarbonate is generated inside the cell (CA II) from carbon dioxide produced in the lumen by CA IV. The net effect is that for every bicarbonate ion filtered, one is returned to the plasma along with sodium, and a proton is secreted into the lumen where it appears as water. Note that filtered bicarbonate passes into the tubular cell as carbon dioxide rather than HCO_3^-, but in quantitative terms, the filtered load of bicarbonate is effectively reabsorbed in the proximal tubule.

Ammonia is a toxic metabolic waste product. The majority of the NH_3 produced each day is ultimately excreted as urea, the remainder being excreted in the urine as NH_4^+. Unusually for an excreted molecule, none of the ammonia that eventually appears in urine is filtered at the glomeruli, as all of it is derived from local generation in, and secretion from, the tubular cells.

Mechanisms for the production of ammonia by the renal tubular cells are crucial as they are responsible for about two-thirds of the bicarbonate that must be generated to replace that lost in buffering the ECF. Ammonia, being a small uncharged molecule, is able to cross cell membranes by simple diffusion. However, the pK_a for protonation of NH_3 is approximately 9.2, so at typical cellular and extracellular pH most (>99%) is present as NH_4^+;

$$NH_3 + H^+ \rightleftarrows NH_4^+$$

The ammonium ion is much less likely than ammonia NH_3 to diffuse across lipid bilayers, so its transport requires specific translocase proteins to mediate its secretion from tubular cells to the lumen of the nephron. Significantly, ammonium can compete for transport with potassium or hydrogen ions. A particularly important example of just such a translocase is designated NHE-3 (Na-hydrogen exchanger-3).

The ammonia in the proximal part of the renal nephron is derived from the amino acid glutamine in a two-step reaction catalysed by the mitochondrial enzymes glutaminase and glutamate dehydrogenase (GlDH):

$$\text{Glutamine} \xrightarrow{\text{Glutaminase}} \text{glutamate} + NH_3 \xrightarrow{\text{GlDH}} \text{oxoglutarate} + NH_3$$

Glutamine, which may itself have been synthesised in the liver, enters the proximal tubular cell from both the plasma and the glomerular filtrate.

Under normal conditions of acid-base homeostasis, about 70% of the ammonia generated by glutaminase/glutamate dehydrogenase enters the lumen and the remaining 30% diffuses directly into the renal veins allowing transport to the liver, where it may either be used to synthesise urea or converted back into glutamine which in turn is recycled to the kidney (see Figure 3.4). Urea synthesis is, however, a net consumer of bicarbonate, so reducing the effectiveness of bicarbonate generation:

$$HCO_3^- + NH_3 + 2ATP \longrightarrow \underset{\text{carbamoyl phosphate}}{H_2NCO\text{-}P} + 2ADP + Pi$$

or

$$HCO_3^- + Gln + 2ATP \longrightarrow H_2NCO\text{-}P + 2ADP + Pi + Glu$$

(Glu is converted back to Gln with NH_3)

Glu = glutamate, Gln = glutamine

Carbamoyl phosphate produced in the reaction above enters the Krebs-Henseleit urea cycle.

The complete deamination of glutamine generates two molecules of ammonia but also two molecules of bicarbonate:

$$\text{Gln} \longrightarrow 2\,NH_4^+ + 2HCO_3^- + \text{phosphoenolpyruvate (PEP)}$$
$$\downarrow$$
glucose
via gluconeogenesis

A small proportion of the NH_3 diffuses directly to the lumen, but most is protonated to NH_4^+ and is secreted via NHE-3 in exchange for Na^+ (cation) in parallel with bicarbonate (anion), thus maintaining electrical neutrality.

Secretion of ammonium occurs throughout the length of the nephron, but a significant concentration gradient, generated by extrusion of NH_4^+ from the ascending limb of the loop, exists between the medullary tissue and the lumen of the nephron. As luminal fluid passes through the collecting duct, the gradient facilitates passage of NH_3 into the lumen whereupon it is protonated to produce NH_4^+, the excreted form (Figure 3.5).

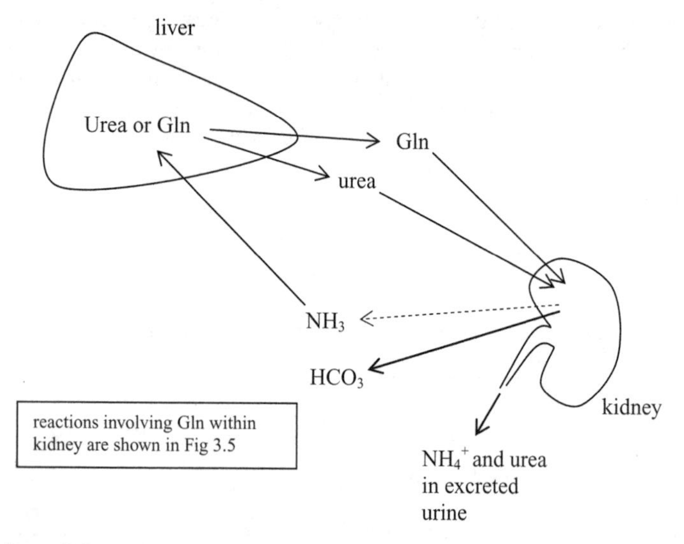

Figure 3.4 Glutamine (Gln) is a carrier of ammonia. The small amount of NH_3 that enters the bloodstream can be utilised within the liver for the formation of either Gln or urea

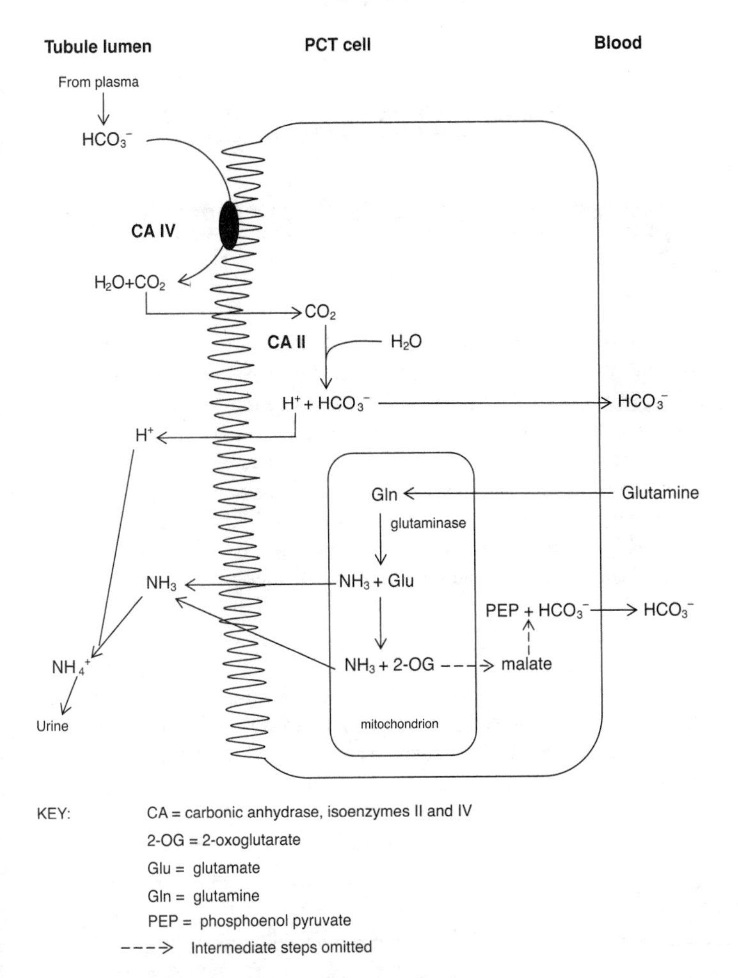

Figure 3.5 Carbonic anhydrase and glutaminase
Deamination of glutamine occurs within mitochondria. 2-oxoglutarate enters the TCA cycle where malate is produced. Malate exits the mitochondrion and is converted first into oxaloacetate and then phosphoenolpyruvate and CO_2, which becomes hydrated to HCO_3^-. PEP is a substrate for gluconeogenesis.

SECTION 3.vii

Respiration: gas pressures and breathing

The 'amount' of oxygen or carbon dioxide present in the blood is measured not by its concentration but by its partial pressure, symbol P, i.e. PCO_2 and PO_2. Some older textbooks may use pCO_2 and pO_2 respectively to indicate the gas pressures. You may also come across abbreviations $PaCO_2$ and PaO_2, where 'a' indicates that the value is for arterial blood, whereas P_A symbolises gas pressure in the alveoli (singular = alveolus), the small terminal air sacs within the lungs. The abbreviations PCO_2 and PO_2 will mostly be used in the following account.

PO_2 values are determined by the atmospheric PO_2, the ventilation (i.e. ease of gas entry) and gas exchange within the lungs. PCO_2 of the blood is determined by the balance between CO_2 production (from tissue respiration) and excretion (ventilation, i.e. rate and depth of breathing). In fact, the rate and depth of respiration are determined mainly by the PCO_2. Special cells located in certain blood vessels, and in contact with the CSF within the central nervous system, act as CO_2 'monitors'. When the PCO_2 rises, respiration is stimulated and vice versa. We are more tolerant of mild hypoxia (low PO_2) than we are of hypercapnia (high PCO_2), as illustrated in Figure 3.6.

Gas exchange in the lung

Gas pressure ($\sim$ concentration) gradients of oxygen and carbon dioxide determine their diffusion between the alveolar spaces and the venous blood arriving at the lungs.

The barriers to diffusion are the normally thin membranes of the alveolus and the blood capillary (Figure 3.7a). Certain diseases of the

---

Essential Fluid, Electrolyte and pH Homeostasis, First Edition. Gillian Cockerill and Stephen Reed.

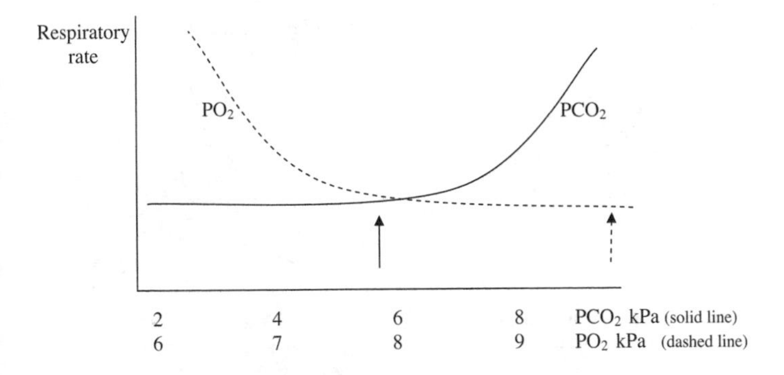

The solid arrow indicates the upper limit of normal for PCO_2.
The dashed line indicates the lower limit of normal for PO_2.

Figure 3.6 Influence of arterial gas pressures on respiratory rate. A rising PCO_2 is a relatively more potent stimulus of respiration than is a falling PO_2
Note: The plot shown for PO_2 is representative as the actual relationship between PO_2 and respiratory rate is affected by blood pH and PCO_2.

lung may either restrict the entry of oxygen and expulsion of carbon dioxide (Figure 3.7b), or impede diffusion across the membranes (Figure 3.7c). Not all of the alveolar sacs are in contact with a capillary, so are in effect 'dead-ends' for oxygen uptake and do not allow carbon dioxide diffusion out of the blood (Figure 3.7d).

As described in the previous section, the kidney deals with approximately 75 mmol of organic metabolic acids each day, but a much larger quantity of acid as CO_2 is removed via the lungs. Cellular respiration utilises oxygen and generates carbon dioxide and ATP to 'drive' metabolism. The generated CO_2 gas diffuses across cell

Table 3.1 Gas pressures. Atmospheric air at sea-level $PO_2 \sim 20$ kPa (kPa = kilopascal)

	Alveoli	Arterial blood	Venous blood
PO_2	14 kPa	13.5 kPa (~85–101 mmHg)	6.5 kPa
PCO_2	6.0 kPa	5.3 kPa (~35–45 mmHg)	6.2 kPa

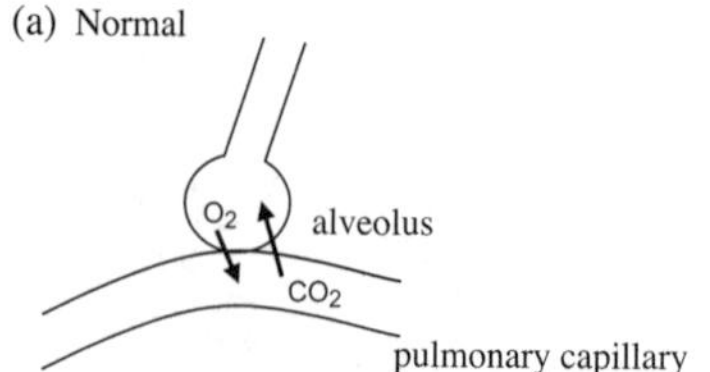

(b) Narrowing of airway or obstruction: impaired gas exchange

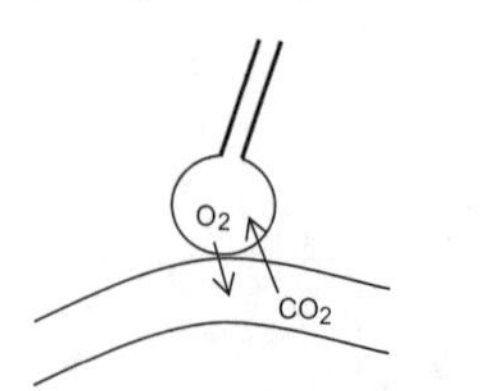

(c) Damage or inflammation to alveolus: impaired gas exchange

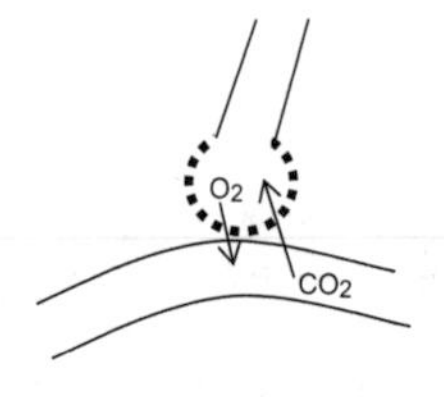

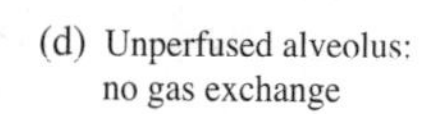

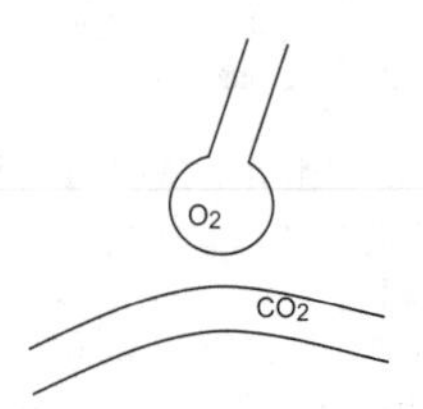

Figure 3.7 Gas diffusion in the lungs

membranes and into the bloodstream where any or all of the following may occur;

(a) CO_2 dissolves in plasma water: $CO_2 \rightleftarrows dCO_2$ ('dissolved CO_2')
In fact, very little of the carbon dioxide dissolves in plasma water: the vast majority passes into the red cells and is hydrated by carbonic anhydrase as shown in (b) below.

(b) spontaneous (i.e. non-enzymatic) hydration forming carbonic acid:

$$CO_2 + H_2O \longrightarrow H_2CO_3$$

(compare with the carbonic anhydrase reaction shown in Section 3.vi)

(c) the carbonic acid so produced may dissociate:

$$H_2CO_3 \longrightarrow H^+ + HCO_3^-$$

(d) further dissociation to carbonate.

$$HCO_3^- \longrightarrow H^+ + CO_3$$

Thus, respiratory carbon dioxide can occur in up to four forms within blood.

The formation of carbonic acid as shown above is central to our understanding of acid-base balance because by excreting CO_2 in expired breath, the lungs effectively rid the blood of acid:

$$\underset{\text{from tissue metabolism}}{\underset{\uparrow}{CO_2}} + H_2O \longrightarrow \underset{\text{bloodstream}}{\underset{\uparrow}{H_2CO_3}} \longrightarrow H_2O + \underset{\text{expired on breath}}{\underset{\nwarrow}{CO_2}}$$

SECTION 3.viii

The role of red cells: gas carriage by haemoglobin

Despite their apparent structural simplicity, lacking as they do any internal organelles, erythrocytes are involved not only with gas carriage, but also have an important role to play in acid-base balance.

Haemoglobin (Hb) transports oxygen from the lungs to the tissues (oxyhaemoglobin, oxyHb), and also transports CO_2 from the tissues to the lungs for excretion (deoxyHb). When red cells leave the lungs *en route* to tissues, they are approximately 99% saturated with oxygen (see Figure 3.8) at a PO_2 of about 13 kPa; when the cells return to the lungs they are approximately 75% saturated with oxygen, at a PO_2 of approximately 5.2 kPa. The term deoxyHb should not therefore be interpreted as haemoglobin which is *totally* without oxygen. These oxygen saturation values suggest that all four haem groups of each haemoglobin molecule carry oxygen away from the lungs, but on return three of the four are still oxygen-bound. If, however, oxygen delivery has to be increased to meet unusual physiological demands, such as vigorous exercise, Hb returning to the lungs may be, say, 60% saturated. The sigmoidal shape of the oxygen dissociation curve (ODC) (Figure 3.8) indicates the allosteric nature of O_2-Hb binding. Oxygen release from haemoglobin responds sensitively to situations in which blood PO_2 is unusually low, as might be the case in times of increased oxygen demand by tissues or if oxygen availability is limited.

Various factors including pH, PCO_2, changes in body temperature and the concentration of a small allosteric modifier molecule called 2,3 *bis* phosphoglycerate (2,3-BPG) will affect the position of the ODC. A shift to the left indicates that oxygen is given up more easily by haemoglobin, so increasing oxygenation of the tissues. A slight fall

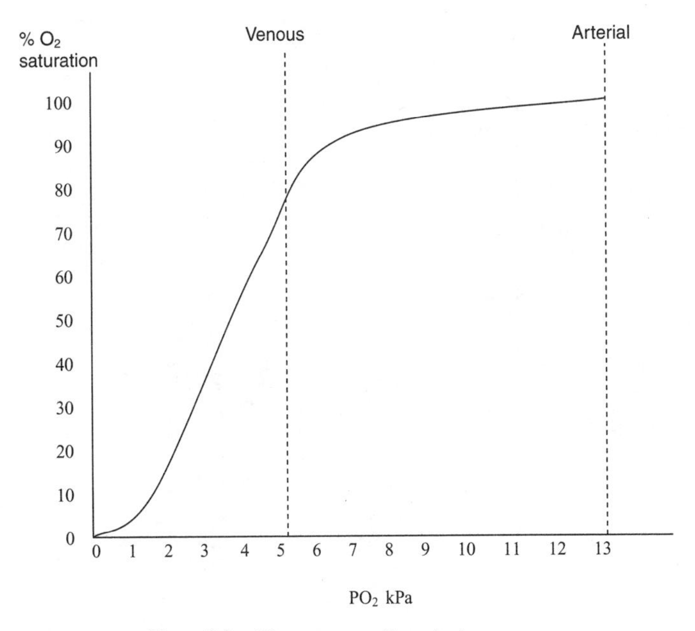

Figure 3.8 The oxygen dissociation curve

in blood pH linked with an increase in PCO_2 would signal accelerated tissue oxidative metabolism requiring increased oxygen delivery, and therefore easier release from haemoglobin. Furthermore, deoxygenated haemoglobin is a stronger base than is oxyhaemoglobin, so any fall in pH may be efficiently buffered by conversion of oxy to deoxyhaemoglobin. If tissue oxygen debt is more prolonged, due for example to low environmental oxygen tension or anaemia, the body's response is to (a) increase the rate of red cell formation and release from the bone marrow, and (b) to increase the concentration of 2,3-BPG in the red cells to promote oxygen release from haemoglobin. Production of 2,3-BPG is via the Rapoport-Leubering shunt, which diverts some 1,3-*bis* phosphoglycerate (1,3-BPG) from glycolysis (see Figure 3.9).

Respiring tissue cells produce carbon dioxide as a waste product; this needs to be removed from the body. Although CO_2 gas is only poorly soluble in water, it will dissolve slightly in plasma to form carbonic acid (H_2CO_3) and dissolved CO_2. A much greater amount of CO_2 is removed by red blood cells (RBCs). When an RBC is in

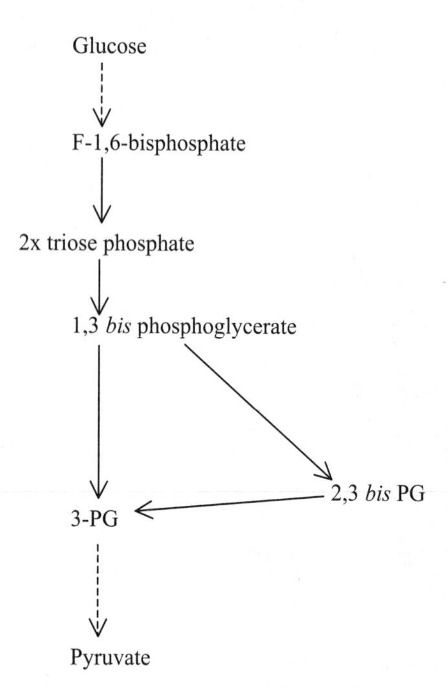

Figure 3.9 2,3-BPG synthesis via the Rapoport-Leubering shunt

a tissue capillary, there exists across the cell membrane gas diffusion gradients for O_2 (higher inside the RBC than in the plasma) and for CO_2 (higher in the plasma than in the RBC). Consequently, the gases diffuse down their respective gradients; O_2 leaves the RBC as CO_2 enters. The CO_2 that enters the RBC may either:

(i) combine with CO_2 from respiring cells more efficiently, forming a carbamino protein (CO_2 reacts with amino groups in the Hb):

$$\text{R-NH}_2 + \text{CO}_2 \rightarrow \text{R-NHCOO}$$

or

(ii) be hydrated to H_2CO_3 by the action of carbonic anhydrase.

Carbonic anhydrase (CA) is an important enzyme because it is found in both the renal tubular cells (Section 3.vi) and also in red blood cells. Recall the key equation:

$$CO_2 + H_2O \xrightarrow{CA} H_2CO_3 \rightarrow H^+ + HCO_3^-$$

The CA accelerates the formation of H_2CO_3, which then dissociates; the proton so produced is buffered by the haemoglobin.

Haemoglobin (Hb) as a buffer and the Bohr effect: deoxyhaemoglobin is a stronger base than oxyhaemoglobin

Haemoglobin is a chemically efficient buffer because its overall isoelectric pH (pI), a reflection of pK_a values of individual amino acid residues, is within the physiological range due in large part to the histidine content. Also haemoglobin is present at high concentrations in whole blood (150 g/L), so this protein offers a high buffer capacity. In addition, the tendency of the protein to accept or donate protons is determined by the extent of oxygenation (Figure 3.10).

When fully oxygenated oxyhaemoglobin ($Hb(O_2)_4$) carried in the arterial supply 'gives up' O_2 to the tissues forming deoxyhaemoglobin ($Hb(O_2)_3$) it undergoes a conformational change which enables the protein to accept, and thereby to buffer, protons generated from the dissociation of H_2CO_3 formed via the carbonic anhydrase reaction.

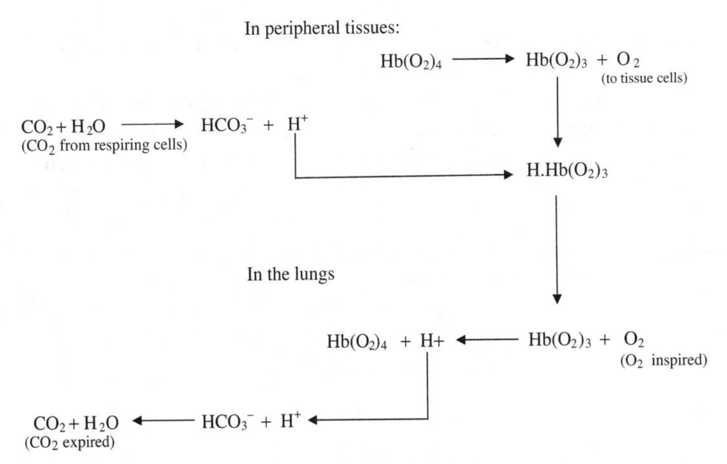

Figure 3.10 The Bohr effect: deoxyHb is a stronger base than oxyHb

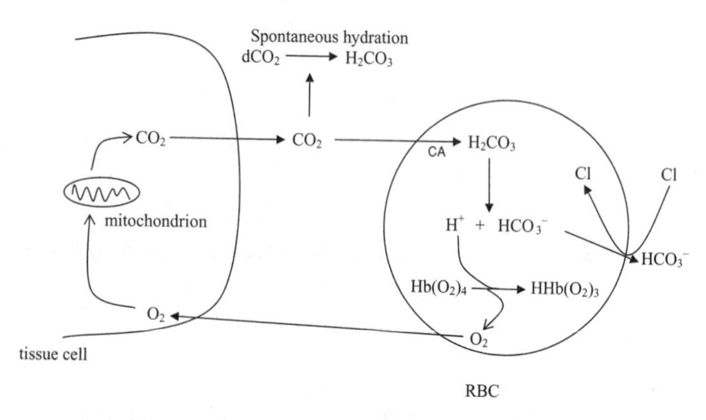

Carbonic anhydrase (CA) in RBC hydrates CO_2;
Hb buffers a proton;
HCO_3^- is exchanged for Cl across the RBC membrane (chloride shift).

Figure 3.11 The role of red blood cells (RBC)

Furthermore, there is an exchange of bicarbonate for chloride (chloride shift) for every H^+ which is buffered. These reactions are illustrated in Figure 3.11. The bicarbonate ion added to the plasma as part of the chloride shift helps to resist the change in pH which would occur as the PCO_2 rises with the spontaneous formation of H_2CO_3. Even so, the net addition of CO_2 to the blood results in venous blood having a higher PCO_2 than arterial blood.

On its return to the lungs, the opposite effects occur and as the haemoglobin becomes once again fully oxygenated, the buffered proton is released and used by carbonic anhydrase to generate a molecule of CO_2.

SECTION 3.ix

Self-assessment exercise 3.2

1. Why is the PO_2 in the alveoli lower than the atmospheric value?
2. Why does venous blood have a lower pH than arterial blood?
3. The ODC shows that the binding between oxygen and the Hb is allosteric. Carbon monoxide is an irreversible competitive inhibitor of oxygen binding. Sketch a diagram to show the effect CO would have on the oxygen dissociation curve. (You may wish to revise some basic protein and enzyme chemistry to answer this question.)
4. What physiological advantage can athletes gain by training at high altitude for a period of time prior to a major sporting event?
5. Calculate the ratio of NH_4^+ to NH_3 at pH 6.85 (ICF pH) given that the pK_a for proton dissociation is 9.2.

Check your answers before continuing.

Essential Fluid, Electrolyte and pH Homeostasis, First Edition. Gillian Cockerill and Stephen Reed.

SECTION 3.x

The liver and gastrointestinal tract in acid-base homeostasis

The kidneys and the lungs control the ratio of bicarbonate to carbonic acid and thus maintain a normal blood pH, but all tissues influence overall acid-base homeostasis by virtue of the fact that they produce metabolic acid and carbon dioxide. The liver and the gastrointestinal systems, however, have a particular role to play in proton turnover.

Digestion of food begins in the stomach. Gastric parietal cells secrete hydrochloric acid (HCl) in response to the anticipation of food or its presence in the stomach. Protons are generated by carbonic anhydrase and secreted via a pump mechanism; chloride is transported along with the proton to maintain electrical neutrality, and the bicarbonate generated by carbonic anhydrase is added to the plasma, a process which has been called the 'alkaline tide'. However, this apparent gain in base is counteracted by the pancreas which secretes a bicarbonate-rich fluid into the duodenum. For each HCO_3^- ion which is secreted into the small intestine, a proton is added to the blood. Carbonic anhydrase is again the enzyme responsible for proton and bicarbonate generation.

In health, most of the gut fluids secreted into the gut are reabsorbed and overall there is only a small net loss of base (about 20 mmol) each day. It is easy to accept that vomiting (loss of HCl) or diarrhoea (loss of alkaline fluid) should be associated with possibly serious acid-base upsets. Such an intuitive interpretation would be challenged by some physiologists whose views conform to the 'quantitative' model

Essential Fluid, Electrolyte and pH Homeostasis, First Edition. Gillian Cockerill and Stephen Reed.

By using lactate as a substrate for gluconeogenesis, the liver effectively removes a metabolic acid from the circulation.

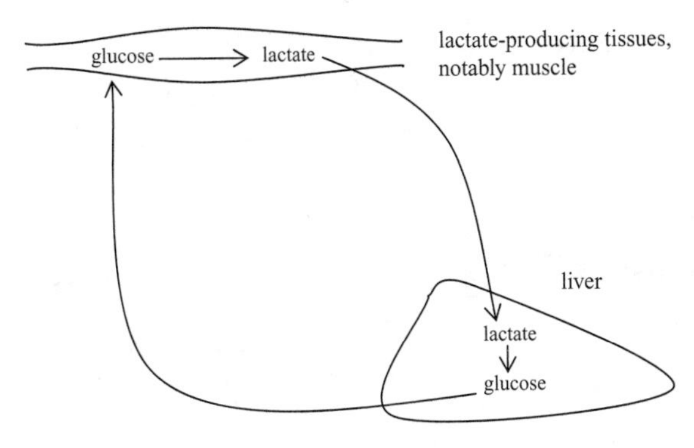

Figure 3.12 Cori cycle

of acid-base imbalance developed by Peter Stewart, as described in the following Section. That theory would explain pH disturbances in diarrhoea or vomiting as the result of loss, via the gut, of electrolyte-containing fluids rather than loss of protons or bicarbonate ions *per se*.

Despite its central role in metabolism, the liver is often overlooked as a key player in acid-base balance. For example, an estimated 1 to 1.5 moles of lactate (= lactic acid) enter the blood each day as a result of the incomplete aerobic oxidation of glucose in tissues. Quantitatively, nearly all of the lactate is removed by the liver and diverted back into carbohydrate pathways by the process of gluconeogenesis, which not only 'removes' a metabolic acid but also generates bicarbonate (Figure 3.12, see also Figure 3.5).

Metabolism of amino acids in the liver generates ammonia which may be used to synthesise glutamine or, if combined with bicarbonate in a reaction catalysed by carbamoyl phosphate synthetase (CPS), undergoes further metabolism in the synthesis of urea, a process that itself produces protons. When the pH of body fluids drops (acidosis) the synthesis of glutamine is accelerated, an important process given that glutamine delivers NH_3 to the kidney for excretion as NH_4^+, whilst urea synthesis is slowed (so fewer protons are generated). In alkalosis when body fluid pH rises, the reverse adjustments are made. The liver

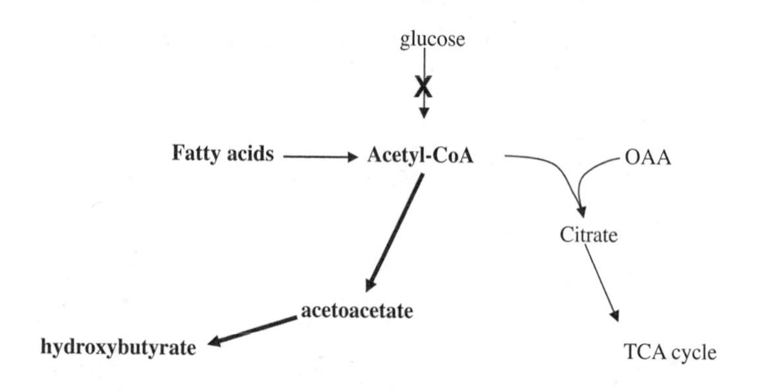

OAA = oxaloacetate

Acetoacetate and hydroxybutyrate are oxo-acids

If glucose is effectively unavailable for energy generation, fatty acids are used instead, but these produce oxo-acids as byproducts.

Figure 3.13 Ketogenesis

is also the site of the oxidation of sulphur-containing amino acids, a process that produces sulphuric acid. During periods of starvation or in certain pathologies, the liver generates oxo-acids, notably hydroxybutyrate and acetoacetate collectively known more commonly as 'ketone bodies', from fat metabolism (Figure 3.13). Even in health, synthesis or catabolism of fatty acids contributes to overall proton turnover.

Finally, the liver is responsible for the synthesis of nearly all of the plasma proteins. Despite its relatively low molar concentration (0.5–0.7 mmol/L), albumin is a particularly important contributor to the buffering capacity of blood, and either hypo- or hyperalbuminaemia can result in an acid-base disturbance.

SECTION 3.xi

The 'traditional' versus the 'modern' view of acid-base homeostasis

At several points in the discussion so far we have focused upon the so-called 'traditional' model of acid-base homeostasis. In this theory there is a perceived central role for the carbonic acid/bicarbonate buffer system and its interpretation through the Henderson-Hasselbalch equation. Thus, the maintenance of a 20:1 ratio of bicarbonate (the 'metabolic' component) to carbonic acid (as PCO_2, the 'respiratory' component) is pivotal. Stated simply, in this model any changes in plasma bicarbonate concentration are seen to be the mechanistic *cause* of a metabolic acid-base disturbance, and changes in PCO_2 are the *cause* of respiratory upsets.

This traditional model was developed primarily from the work of Siggaard-Anderson and Astrup (amongst others) in Scandinavia during the 1950s and 1960s. This interpretation, as described in all textbooks, is intuitively easy to comprehend in simple terms of the concentrations of acid (protons and carbonic acid) and base (bicarbonate, with 'back-up' from phosphate and protein), and has served fairly well in clinical practice for decades. In this model, approximately a third of the buffering capacity within the ICF is ascribed to bicarbonate and the remainder to phosphate and protein; in the ECF, 85% of buffering capacity is due to HCO_3^- and only 15% due to protein and phosphate. Addition of acid to the body fluids 'titrates' bicarbonate-forming carbon dioxide and water, and the bicarbonate is regenerated via renal mechanisms and via metabolic reactions that consume protons:

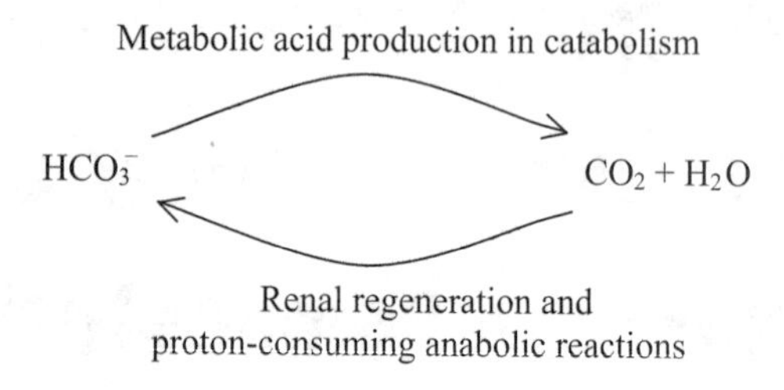

It is noteworthy that cellular acidosis has direct effects on metabolic pathways: glycolysis is slowed by inhibition of phosphofructokinase, the urea cycle which is a net consumer of bicarbonate is slowed, but ammonium ion formation in the nephron (Figure 3.5 in Section 3.vi) is promoted. Thus, bicarbonate consumption and generation are focal events in maintaining a normal pH within body fluids.

This traditional view has, however, been challenged, and in the early 1980s a different approach to the understanding of the underlying mechanisms associated with changes in blood pH was developed by Peter Stewart in the United States. There followed a period of scientific argument which became known as the 'Great transatlantic acid-base debate'; to some extent the debate is still not won.

The so-called 'modern', 'alternative' or 'quantitative' interpretation developed by Stewart identifies three *independent* chemical factors, namely the strong ion difference (SID), the PCO_2 and the total concentration of weak acids (A_{TOT}; mostly protein and a smaller contribution from phosphate), to explain mechanistically how acid-base disturbances arise. Bicarbonate and proton concentrations are *dependent* variables whose values change in response to changes in one of the independent factors, and changes in blood pH can only arise due to changes in one of the three independent factors. Put simply, Stewart views changes in blood bicarbonate (a key player in the traditional model) as the *result*, not the *cause*, of acid-base imbalance. In contrast to the 'traditional' theory, which considers the ratio of HCO_3^- to PCO_2 to be the defining factor in acid-base disturbances, Stewart argues that PCO_2 and HCO_3^- are not independent but mutually *inter*dependent factors. If one changes the other must also change; this can be partly explained by reference to the 'key' equation we met previously:

$$CO_2 + H_2O \rightleftharpoons HCO_3^- + H^+$$

Thus, if PCO_2 rises, so too does HCO_3^- concentration but proportionately less than the corresponding rise in the $[H^+]$; conversely, if PCO_2 falls, so too does HCO_3^-. Thus, to consider PCO_2 and bicarbonate as useful independent reflectors of acid-base status is misleading. The major advantages of the modern theory are that (i) it allows a much better understanding of quantitative changes occurring in acid-base disorders, in that the magnitude of plasma bicarbonate concentration alterations do not necessarily reflect the extent of the acidosis or alkalosis, and (ii) it explicitly takes into account changes in plasma albumin concentration and the other components of the total buffer base. Indeed, even the traditional approach acknowledged that in order to attempt to quantify the severity of acidosis, it is necessary to know the concentrations of all major buffer anions.

Stewart's approach was to apply laws of physical chemistry to the ionic equilibria and dissociations occurring in blood, and to analyse these in a very strict and complex mathematical fashion. Indeed, in a seminal paper on the topic,[1] Stewart acknowledged that the complexity of the mathematics effectively prevented the development of the theory until the advent of computers. The fundamental ideas proposed by Stewart require us to re-think our understanding of the Lowry-Bronsted concept of acids and bases as proton donors or proton acceptors respectively; rather, we need to see acids and bases as compounds that modify the dissociation of water as defined by K_w:

$$[H^+] \times [OH^-] = K_w \times [H_2O]$$

Given the very high (55.5 molar) and essentially constant concentration of water, the $[H_2O]$ term can be ignored because it does not change significantly, so the equation simplifies to:

$$[H^+] \times [OH^-] = K'_w$$

but the dissociation of water liberates equal numbers of protons and hydroxyl ions,

$$\text{i.e. } [H^+] = [OH^-]$$

[1]Stewart, P.A. (1983) Modern quantitative acid-base chemistry. *Can. J. Physiol. Pharmacol.* **61**: 1444–1461.

therefore we can say that $[H^+]^2 = K'_w$

and thus, $[H^+] = \sqrt{K'_w}$

where in pure water at 37°C $K'_w = 4.4 \times 10^{-14}$

A solution is said to be acidic if $[H^+] > \sqrt{K'_w}$

$$\ldots\text{and basic if } [H^+] < \sqrt{K'_w}$$

However, and this is the crucial point, the addition to pure water of compounds that strongly dissociate (e.g. NaCl) will modify the dissociation of water and therefore the $[H^+]$ in a way that is dependent upon the difference in strong ion concentration:

$$[H^+] \propto [Na] - [Cl]$$

The numerical value for $[Na] - [Cl]$, Stewart called the 'strong ion difference' (SID).

From an analysis of simple solutions containing only one strong electrolyte (NaCl), the theory was developed to include all other strong electrolytes, all weak acids (A_{TOT}) and carbon dioxide (PCO_2) in solution, thus a model of pH changes for both blood and ICF was created which is based upon three independent factors: SID, A_{TOT} and PCO_2.

As will be described in more detail in Section 3.xv, acid-base disturbances are broadly classified as 'respiratory' or 'non-respiratory' (also called metabolic) in origin. There is no real disagreement between the traditional and modern schools of thought on acid-base disturbances which have a respiratory origin; the PCO_2 is a good measure of the severity of the problem. The argument over the assessment of the metabolic component is, however, acute. The modern approach uses SID, but even the traditionalists recognise the inadequacy in the use of bicarbonate alone to assess the metabolic insult. In 1957 a derived parameter called Standard Bicarbonate (SB) was introduced to act as a reference point. This is defined as the bicarbonate concentration which would be found at 37°C when $PCO_2 = 5.3$ kPa and the blood is fully oxygenated. Variation from the 'normal' of 24 mmol/L was taken as an indication of acidosis (low SB) or alkalosis (elevated SB). Although a useful measure, SB does not tell the full story because bicarbonate is only one component, albeit the major one, of the total buffer base, i.e. the total buffer capacity of blood. Thus in 1958 researchers within the traditional school introduced another term, the base excess (BE), to help to assess the overall change in buffer anion concentration which is equivalent to stating the extent of the acid challenge. Base excess (BE) may be defined thus: 'the quantity of acid (or base) which must be added *in vitro* to one litre of fully oxygenated whole blood to achieve a pH of 7.40 at body temperature and with a PCO_2 of 5.3 kPa'. The value may be positive, i.e. a real *excess* of base, or negative, sometimes called a base deficit. In practice, whole blood is equilibrated with CO_2 at 5.3 kPa and the BE is calculated by the blood gas analysers commonly used in wards and clinical departments.

The purpose of the BE is to provide a numerical value that reflects only 'metabolic' causes affecting blood pH. By adjusting the sample to a PCO_2 of 5.3 kPa, any respiratory cause of abnormal pH is effectively eliminated. The reference range for BE is +2 to −2 mmol/L; values less than −2 indicate an acidosis and values greater than +2 indicate an alkalosis. The practical value of base deficit is that it allows clinicians to estimate how much bicarbonate must be added to intravenous infusions in order to 'titrate' a patient's blood back to a normal pH, and so attempts to give a quantitative estimate of the severity of an acidosis or alkalosis. Sometime later, the Standard Base Excess (SBE) was introduced to be an even better reflection of deranged metabolic control *in vivo* (compare with the fact that BE is defined *in vitro*). The SBE assumes a haemoglobin concentration of 5 g/dL, which gives a better reflection of the change to the ECF as a whole rather than just the blood.

SECTION 3.xii

Stewart's three independent factors

The outcome of Stewart's mathematical analysis of pH regulation in body fluids (Section 3.xi) was the development of a complex fourth-order polynomial equation showing $[H^+]$ as a function of $K_{w,}$ strong ion difference, PCO_2 and A_{TOT}. Details of the mathematical derivation will not be given here, but the three independent factors are described.

(1) Strong ion difference (SID)

This is *conceptually* (not mathematically) similar to the anion gap (Section 1.x). The term 'strong' relates to the extent of dissociation, not of an acid in this case, but of a salt or a conjugate. Strong ions are fully dissociated from their counter-ions. For example:

$NaCl \rightarrow Na^+ + Cl^-$ whereas weak ions are poorly dissociated from their counter-ion, e.g. H_2CO_3, which only partially dissociates into H^+ and HCO_3^- ions. Strictly speaking, it would probably be more accurate to refer to 'strong *conjugates*' rather than simply 'strong *ions*', but the term is embedded within the literature.

Strong ions include: Na^+, K^+, Ca^{2+}, Mg^{2+} and Cl^-. $Lactate^-$, although not as strong as, say, Cl^-, is well dissociated in blood under physiological conditions, so it is classified as strong for the purposes of the discussion. The pK_a for lactic acid is 3.8, so the ratio of lactate anion to the undissociated conjugate acid in blood at pH 7.40 is of the order of 1000:1.

Weak ions include: protein (the amino and carboxyl groups in proteins are weakly ionised), phosphate ($H_2PO_4^-$, $pK_a = 6.8$) and bicarbonate derived from carbonic acid ($pK_a = 6.1$). Because bicarbonate

Essential Fluid, Electrolyte and pH Homeostasis, First Edition. Gillian Cockerill and Stephen Reed.

is classified as a weak anion, its concentration is seen to be *dependent upon SID*, i.e. determined by SID, so cannot itself be responsible for initiating changes in ECF pH. The concept of SID was originally developed with just Na and Cl in aqueous solution:

$$SID = \{[Na^+] + [H^+]\} - \{[Cl] + [OH^-]\}$$

In blood the strong cations exceed the strong anions by approximately 40 mmol/L (or, more precisely, 40 mEq/L because we are dealing with charge, not ion concentration).

$$SID = ([Na^+] + [K^+] + [Mg^{2+}] + [Ca^{2+}]) - ([Cl^-] + [lactate^-] + [SO_4{}^{2-}]) \sim 40\ mmol/L \equiv 40\ mEq/L$$

Strictly speaking, the equation above defines the *apparent* SID (SID_{app}), i.e. the value that is apparent from the concentrations of strong ions which may be measured in the laboratory.

In a similar fashion to that we used to derive the anion gap, we can explain and understand SID in terms of the need for the maintenance of electrical neutrality:

$$\text{total cations} = \text{total anions} \quad \text{or} \quad [\text{total cations}] - [\text{total anions}] = 0$$
$$\therefore \text{total (strong cations + weak cations)} = \text{total (strong anions + weak anions)}$$

The contribution of weak cations such as H^+ is quantitatively small, so to a very good approximation:

$$\text{total strong cations} = \text{total (strong anions + weak anions)}$$
$$[\text{total strong cations}] - [\text{total (strong anions + weak anions)}] \sim 0$$
$$\text{thus,} \quad \text{weak anions} = (\text{total strong cations} - \text{total strong anions}) = SID$$

Clearly, SID will be positive if [strong cations] exceed [strong anions] and *vice versa*.

Physiologically, of course, there is no 'ion' (i.e. charge) difference as weak anions such as proteinate$^-$ (mostly albumin), phosphate$^-$ and bicarbonate$^-$ account for the value of 40 mEq/L. In effect, therefore, the SID is an estimate of unmeasured weak anions whose concentrations are determined by SID. The magnitude of the contribution of each of the major weak anions has been quantified thus:

$$SID = [HCO_3{}^-] + \{[Alb](8.0pH - 41)\} + \{[Pi](0.3pH - 0.4)\}$$

where all values for phosphate (Pi), bicarbonate and albumin are in mmol/L.

Typical plasma values are: HCO_3^- = 25 mmol/L; albumin = 0.62 mmol/L; Pi = 1.1 mmol/L.

The reader should note that a large number of formulae such as the one above purporting to 'correct' for various changes in acid-base parameters have been developed; it is probably not necessary to memorise them all.

A critical point mentioned earlier, and assuming Na^+ and K^+ concentrations to be constant, is that it is the concentration of chloride (and possibly other anions such as lactate which are normally present in low concentrations) which *determines* the concentration of bicarbonate in blood. Thus, according to Stewart, it is changes in the SID which better explain the *mechanism* of acid-base imbalance, and blood bicarbonate concentration merely reflects the change in strong ion concentration. In effect, bicarbonate ions may be 'displaced' if the concentration of chloride or other anions, lactate for example, rises and bicarbonate concentration increases when chloride is lost as a result of vomiting. Moreover, in this model, the excretion of NH_4^+ or $H_2PO_4^-$ by the kidney occurs not primarily to remove protons, but rather to change the SID as chloride is lost with NH_4^+ and sodium is lost with $H_2PO_4^-$. In sharp contrast to the traditional theory, the kidney contributes to acid-base homeostasis via regulation of SID, *not* directly via proton and bicarbonate ion secretion or reabsorption.

Finally, and of note, is that part of Stewart's mathematical analysis showed the dependence *of* [H⁺] *on* SID, which in very simplified form is:

$$[H^+] = \frac{K'_w}{SID}$$

This serves to emphasise the relationship between $[H^+]$ from the dissociation of water as a major source, perhaps *the* major source of protons and SID. If we assume that K'_w is a constant, then there is a simple inverse relationship between SID and $[H^+]$. Changes in SID therefore *initiate* changes in ECF pH, so when SID increases, $[H^+]$ decreases (pH rises) and vice versa. A range of so-called 'metabolic' acid-base disturbances associated with, for example, renal dysfunction, diabetes, gastrointestinal pathology and drug/toxin intake can be explained in terms of abnormalities in SID rather than alteration in the plasma bicarbonate concentration and thus the 20:1 ratio defined by the Henderson-Hasselbalch equation.

(2) Carbon dioxide

As described earlier (Section 3.vii), the CO_2 content of the blood is determined by the balance between production (tissue respiration) and excretion (respiratory rate × depth of breathing), that is, PCO_2 is independent of any other acid-base factor. As CO_2 passes into the bloodstream, a proportion of it dissolves in plasma water forming carbonic acid, H_2CO_3. The solubility coefficient for CO_2 (0.225 when PCO_2 is expressed in kPa, or 0.03 when PCO_2 is given in old-fashioned units mmHg) varies with temperature, ionic strength and the other solute concentrations. In this respect, the two models of acid-base homeostasis do not differ, and so interpretation of respiratory-related pH changes does not cause any difficulty; raised PCO_2 leads to acidosis and reduced PCO_2 is associated with alkalosis.

(3) Total weak acid concentration, A_{TOT}

Here, HA represents the total concentration of all *non-volatile* weak acids in blood.

$$[A_{TOT}] = [HA] + [A^-]$$

In reality, this equates to proteins, mainly albumin, and phosphate:

$$[A_{TOT}] = [alb^{12-}] + [phosphate^{2-}]$$

Working from the principle of conservation of mass, Stewart argued that assuming the weak acids present in blood take no part in other reactions, the sum of [HA] and $[A^-]$ will always remain constant. The practical importance of A_{TOT} is that it explains why changes in plasma albumin concentration have an impact on blood pH, which the traditional bicarbonate-based model cannot explain. If plasma albumin concentration falls, blood pH rises. Because the normal concentration of phosphate in ECF is very low (~1 mmol/L) hypophosphataemia has little impact on pH, but if, as for example in renal failure, plasma phosphate concentration rises, a fall in pH would be expected.

Effective SID and SIG

The apparent SID (SID_{app}) was defined above (page 263), but two other calculated parameters are required to interpret acid-base data fully using the Stewart approach: (i) the effective SID (SID_{eff}), and (ii) the strong ion gap (SIG).

SID_{eff} is a calculated parameter that is based on measurement of $[HCO_3^-]$ and estimated anionic charge contributions of protein (x−) and phosphate (y−), both of which are pH-dependent.

$$\text{Assuming,} \quad \text{SID mEq/L} - A_{TOT} \sim 0$$
$$\therefore \quad SID_{eff} \sim [alb^{x-}] - [phosphate^{y-}] + [HCO_3^-]$$

The presence of strong ions other than those given in the SID_{app} equation will also influence the value of SID_{eff}. From this, it follows that:

$$SIG = SID_{app} - SID_{eff}$$

and that the numerical value of SIG is due to unmeasured strong ions. SIG is positive if there are unmeasured strong anions present; conversely, SIG is negative if there are unmeasured strong cations present. In health, SIG = 0; SIG is positive if the unmeasured anion concentration exceeds the unmeasured cation concentration (typical of acidosis).

Application of the alternative theory allows interpretation of respiratory disorders (via the PCO_2) and metabolic conditions. For example, renal dysfunction affecting the kidneys' ability to regulate the SID via Na^+ and Cl^- handling, in liver disease when too little lactate is being metabolised or too little albumin is being synthesised, and also in gut disorders when loss or reabsorption of strong ions within digestive secretions is compromised; significantly, the traditional model does not deal with these.

In summary, the traditional theory is conceptually easy to understand in terms of well-established ideas of acids and bases, and proponents of this approach maintain that correct diagnoses can be achieved with $[HCO_3^-]$, PCO_2, supplemented with the anion gap. Furthermore, the conceptual difficulty in the number of parameters used and complication of the mathematics required to deal with so many individual components undermines the modern approach. A major criticism of the Stewart approach, however, is its reliance on the combined use of several parameters, the measurements of which are all subject to error. Even small analytical errors are cumulative, so the potential for there to be a significant overall error when considering possibly 6 or 8 values together cannot be discounted. Supporters of the Stewart model maintain that it provides a better insight into the mechanisms of acid-base disturbance and is more reliable, especially in complex cases as will be encountered in intensive care situations or where changes in plasma albumin concentration have occurred.

Summarising, the pH of body fluids falls if SID falls or if there are increases in A_{TOT}, protein concentration or chloride concentration.

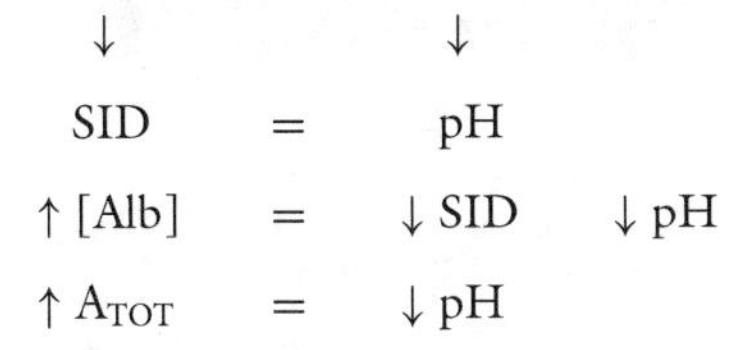

The debate in the scientific literature is still ongoing, and in practice, it is likely that both paradigms will be used pragmatically as particular situations demand.

SECTION 3.xiii

Laboratory measurement of pH, PCO_2 and bicarbonate

Gas probes

In this section we will discuss the importance of measuring the partial pressures of carbon dioxide (PCO_2) and oxygen (PO_2). Both gases are quantified by the use of electrometric devices. Section 2.xiv describes briefly how ion selective electrodes (ISEs) are used to estimate the concentrations of electrolytes in plasma. The same principles are utilised in the laboratory to measure the pH and the PCO_2, but here the glass membrane that actually senses the particular ion of interest is responsive to hydrogen ions. The PCO_2 electrode (named after its inventor, Severinghaus) is a modified pH probe. Surrounding the H^+ sensitive glass membrane is a thin layer of bicarbonate/carbonic acid sealed within a gas permeable plastic film. When in contact with the blood sample, CO_2 diffuses across the plastic film and disturbs the equilibrium between H_2CO_3 and HCO_3^-, causing a local change in the $[H^+]$. It is this local pH change which is detected by the electrode and related to the PCO_2. This is a practical application of the 'key' equation we have met in a physiological context previously:

$$CO_2 + H_2O \longrightarrow H_2CO_3 \longrightarrow HCO_3^- + H^+$$

The change in proton concentration (measured as pH) is detected by the ISE, but the result is given as partial pressure of CO_2.

Essential Fluid, Electrolyte and pH Homeostasis, First Edition. Gillian Cockerill and Stephen Reed.

The oxygen electrode (also known as the Clark electrode) uses a platinum cathode and a silver anode in contact with a saturated solution of potassium chloride to estimate the partial pressure of oxygen, PO_2. This device is an amperometric electrode, which means that it measures the current flow (electrons) in a circuit set up between two electrodes that are held at a constant potential difference of 600 mV. The electrodes are covered by a gas-permeable polypropylene or Teflon membrane. Oxygen in the sample diffuses across the membrane and is reduced at the cathode by addition of electrons which are generated at the anode:

$$\text{Ag anode reaction} \qquad 4Ag^+ + 4Cl^- \longrightarrow 4AgCl + 4e^-$$

$$\text{Pt cathode reaction} \qquad O_2 + 4H^+ + 4e^- \longrightarrow 2H_2O$$

The more oxygen that is available for reduction, the greater is the current flow.

The pH electrode (or pH probe) is a well-known and commonly used piece of laboratory equipment. Typically, both the reference electrode and the sensing electrode are made of silver coated with silver chloride, and the reference electrode is immersed in a solution of potassium chloride.

To measure changes in blood pH reliably, the device must be extremely electrically stable. The Nernst equation predicts that a 1 pH unit change (10-fold change in $[H^+]$; strictly speaking H^+ *activity*) would cause a voltage change of 59 mV (see equations 5, 6 and 7 on page 143). Clearly, to monitor changes in blood pH in the physiological range 7.00–7.80 requires an instrument of great precision.

Plasma bicarbonate concentration is usually measured with a two-step enzyme-based method:

$$\text{pyruvate} + HCO_3^- \xrightarrow{\text{PC}} \text{oxaloacetate}$$

$$\text{oxaloacetate} + \text{NAD} + H^+ \xrightarrow{\text{MDH}} \text{malate} + \text{NAD}^+$$

where PC = pyruvate carboxylase, MDH = malate dehydrogenase.

Assuming that HCO_3^- is the limiting concentration, the rate of change of NADH to NAD^+ is a direct measure of the $[HCO_3^-]$.

Some older methods for estimating plasma bicarbonate concentration actually measured the 'total CO_2' (tCO_2) content of plasma. The numerical difference between true or actual $[HCO_3^-]$ and tCO_2 is about 1.5 mmol/L in normal samples but could be greater in certain pathologies. Actual bicarbonate (A.HCO_3) is not usually measured but may be calculated:

$$\text{A.HCO}_3 = \text{tCO}_2 - (\text{dissolved CO}_2 + \text{H}_2\text{CO}_3 + \text{CO}_3 + \text{carbamino-CO}_2)$$

Carbamino-CO_2 is carbon dioxide bound to proteins as amide groups.

Acid-base disturbances

SECTION 3.xiv

Classification of primary changes based on pH and aetiology

Abnormalities in acid-base status often accompany pathology. Specific changes in parameters such as PCO_2, HCO_3^- and SID may be part of the clinical presentation of a wide range of disorders, and the terms used to classify acid-base upsets, as given in this Section, help to define the symptoms of those disorders. The descriptions and explanations of disturbances that follow are based on the traditional carbonic acid/bicarbonate theory using the Henderson-Hasselbalch equation, but reference will be made to the Stewart model as appropriate.

The Henderson-Hasselbalch equation as applied to bicarbonate/carbonic acid, or more accurately PCO_2, states;

$$pH = pK_a + \log_{10}\left\{\frac{[HCO_3^-]}{PCO_2 \times 0.225}\right\}$$

Simplifying this equation emphasises the important relationships:

$$pH \propto [HCO_3^-]$$
$$pH \propto 1/[PCO_2]$$

Primary changes in acid-base status will arise when:

(a) the PCO_2 is altered (increased or decreased) from its normal value of 5.3 kPa (range 4.7–5.9 kPa). Typically, a change of PCO_2 by

Essential Fluid, Electrolyte and pH Homeostasis, First Edition. Gillian Cockerill and Stephen Reed.

1.6 kPa equates to a change in pH of 0.1 unit, assuming $[HCO_3^-]$ remains constant.

or,

(b) the plasma HCO_3^- concentration is altered from the normal value of approximately 25 mmol/L (range 23–28 mmol/L). Bicarbonate is the major contributor to base excess and a change in BE of 5–6 mEq/L will bring about approximately a 0.1 unit change in pH, assuming PCO_2 does not change.

Disturbances that result in an initial change in the PCO_2 are referred to as respiratory disorders. Hypercapnia means the PCO_2 is above the normal range and arises when the production of CO_2 exceeds the ability of the respiratory system to excrete it. Hyperventilation, increased respiratory rate, will reduce the PCO_2.

Disturbances that result in primary changes in plasma $[HCO_3^-]$ are referred to as non-respiratory (also called metabolic disorders). Non-respiratory disorders can arise due to loss of base (bicarbonate) or due to titration of bicarbonate following proton accumulation. Poor oxygen delivery (hypoxia) to tissues, although usually a respiratory problem, causes impaired metabolism and the consequent acid accumulation leads to a metabolic disorder.

Assuming the blood pH is known, we can define precisely the type and primary cause of any acid-base abnormality:

(a) Respiratory acidaemia
(b) Respiratory alkalaemia
(c) Non-respiratory (= metabolic) acidaemia
(d) Non-respiratory (= metabolic) alkalaemia
(e) Mixed disturbance: any combination of two of (a) to (d) above.

By definition, an acidosis would be associated with a negative base excess (or base deficit) if we rely on the traditional model of data interpretation. A differential diagnosis of the cause of the metabolic acidosis can be made according to the anion gap (AG).

$$AG = ([Na^+] + [K^+]) - ([HCO_3^-] + [Cl^-])$$

Any such 'gap' is purely analytical, since physiologically we know that to maintain electrical neutrality it is imperative that 'total positive charge = total negative charge'. The 'gap' is due to unmeasured anions, for example lactate, ketoacids, phosphate, sulphate and proteins.

Assuming the 'normal' AG is less than 15 mEq/L, values for the AG of greater than 20 mEq/L represent a significant abnormality. An increased AG is consistent with accumulation of metabolic organic acids (lactic acidosis, ketoacidosis), ingestion of organic acids or metabolites of ingested acids (salicylate, methanol and other toxins), renal failure (phosphate and sulphate) and rare inborn errors of metabolism. Acidosis with a normal AG is a sign of hyperchloraemia, itself a result of intake or infusion of chloride salts, or loss of bicarbonate via the gut or kidney. Thus, AG plus BE offers a useful insight into the origin of any pathology.

SECTION 3.xv

Overview of mechanisms

Interpretation and classification of acute (i.e. recent onset) disturbances is relatively easy as we need only to consider the relative change in PCO_2 or bicarbonate concentration. However, with time and in chronic disorders the body begins to adapt to the acid-base insult and both bicarbonate concentration and PCO_2 may change, making identification of the primary upset somewhat difficult.

(a) Respiratory imbalances

According to the Henderson-Hasselbalch equation, pH is *directly* proportional to $[HCO_3^-]$ and *inversely* proportional to PCO_2. Carbonic acid concentration is determined by PCO_2, (i.e. $[H_2CO_3] = 0.225 \times PCO_2$), so a decrease in PCO_2 leads to a decrease in $[H_2CO_3]$ and therefore a *rise* in pH.

(i) *Respiratory acidosis*

This condition will occur whenever the lungs expel CO_2 less quickly than it is produced by the tissues as a result of normal metabolism. The result is accumulation of CO_2 in the blood and so a rise in the PCO_2 and thus a fall in blood pH. This may be due to hypoventilation (reduced respiratory rate) or due to damage to the lung tissue preventing normal gaseous exchange. Typical causes include:

- chronic lung diseases;
- damage to the respiratory centre in the brain. The action of breathing is controlled by the respiratory centre so any damage, whether it be physical or dru-induced, will cause a reduction in the frequency and/or depth of breathing (hypoventilation).

Essential Fluid, Electrolyte and pH Homeostasis, First Edition. Gillian Cockerill and Stephen Reed.

(ii) *Respiratory alkalosis*

An alkalosis will occur whenever the lungs expel CO_2 more quickly than it is produced by the tissues as a result of normal metabolism, so the blood pH rises because the PCO_2 falls. Typically this is due to:

- stimulation of the respiratory centre in the brain. The respiratory centre controls the action of breathing so any change in neural stimulation of the respiratory muscles by the respiratory centre will cause an increase in the frequency and/or depth of breathing;
- patients who are being maintained on a mechanical ventilator because their own physiological control of breathing is defective.

(b) Non-respiratory ('metabolic') imbalances

In non-respiratory (more frequently called metabolic) disturbances, the initial alteration in blood chemistry is in the plasma [HCO_3^-]. The Henderson-Hasselbalch equation shows that there is a direct relationship between pH and [HCO_3^-].

(i) *Metabolic acidosis*

In this type of acid-base upset, the concentration of HCO_3^- is too low. This situation can occur for the following reasons:

- overproduction of metabolic acid
- excessive loss of HCO_3^- via the kidneys
- failure to excrete a normal acid load.

Determination of the anion gap (AG, see page 274) is useful when trying to identify the exact cause of a metabolic acidosis (see Figure 3.14). An increase in the AG signifies the presence of an unmeasured anion such as lactate or ketoacids and so would suggest the cause as an overproduction of metabolic acid. A normal AG with low blood pH usually indicates hyperchloraemia; the plasma bicarbonate concentration is low because it has been displaced by chloride ions.

(ii) *Metabolic alkalosis*

In this type of acid-base upset, the concentration of HCO_3^- is too high. There are very few genuine pathological causes of this. Occasionally, patients can induce this sort of condition by inappropriate self-administration of certain drugs.

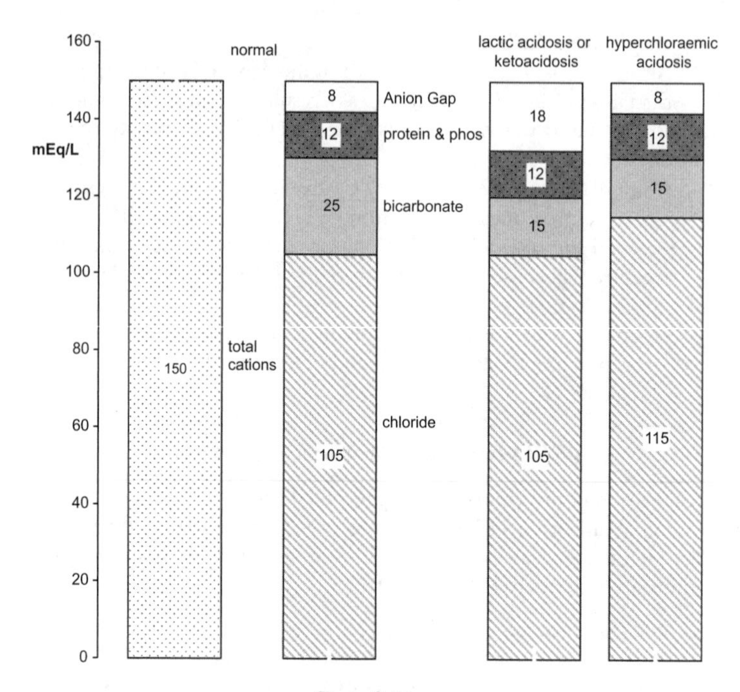

Figure 3.14

(c) Mixed disturbances

Rarely, a patient may present with a mixed disturbance, with two or even three underlying abnormalities being present simultaneously. For example:

(i) a patient on diuretic therapy (metabolic acidosis) who suffers a bout of vomiting (metabolic alkalosis) or diarrhoea (metabolic alkalosis);
(ii) someone who has suffered a cardiac arrest (hypoxia, metabolic acidosis and respiratory acidosis);
(iii) an aspirin overdose (metabolic acidosis and respiratory alkalosis).

SECTION 3.xvi

Physiological correction of primary disturbances

Because an abnormally high or low pH will compromise essential physiological functions, the body responds to a primary alteration in either PCO_2 or $[HCO_3^-]$ by trying to restore the all important base:acid (i.e. $[HCO_3^-]$:$[H_2CO_3]$) *ratio* of 20:1. The *absolute* concentrations of base $[HCO_3^-]$ and acid $[H_2CO_3]$ may be abnormal, but providing that the ratio is 20:1 the blood pH will be normal. According to the Henderson-Hasselbalch equation, given that the pK_a is constant, the blood pH varies as a function of the log of the ratio of the bicarbonate concentration and the carbonic acid concentration.

The ability of the body to respond automatically to an acid-base upset is called '*compensation*'. The importance of the dual physiological control (lungs and kidneys) in regulating the bicarbonate to carbonic acid ratio is now apparent; the kidneys control bicarbonate reabsorption (coupled with proton secretion) whilst the lungs regulate the PCO_2. Adapting the Henderson-Hasselbalch equation to define the physiological control processes, we can say that:

$$\text{pH} \quad \propto \quad \text{kidney function/lung function}$$

For example, in order to compensate for a respiratory alkalaemia, the kidney will reabsorb less HCO_3^-; thus the plasma $[HCO_3^-]$ will fall until the $[HCO_3^-]$:$[H_2CO_3]$ ratio is returned to 20:1. In effect, the body is imposing upon itself a metabolic acidosis. Also, in order to compensate for a non-respiratory acidaemia (e.g. a diabetic coma) the lungs would expel CO_2 more quickly to reduce PCO_2 until the base:acid ratio is returned to 20:1 (see Table 3.1). This hyperventilation will reduce PCO_2 (an imposed respiratory alkalosis). The

Essential Fluid, Electrolyte and pH Homeostasis, First Edition. Gillian Cockerill and Stephen Reed.

Table 3.1 Physiological compensation to primary acid-base insults

Primary change	Compensatory change
Respiratory acidosis	Induced non-respiratory alkalosis
PCO_2 rises	Increase [bicarbonate] by enhanced generation in the nephron
Non-respiratory acidosis	Induced respiratory alkalosis
$[HCO_3^-]$ falls	Hyperventilation to reduce PCO_2
Respiratory alkalosis	Induced non-respiratory acidosis
PCO_2 falls	Reduce [bicarbonate] by decreased reabsorption/generation in nephron
Non-respiratory alkalosis	Induced respiratory acidosis
$[HCO_3^-]$ rises	Hypoventilation to increase PCO_2

responses of the respiratory system to changes in pH are, in health, virtually instantaneous, whereas the renal compensation often takes a day or two to become effective.

In practice, compensation may not be 'perfect' and the blood pH may be at the limit of normal, but at least the initial challenge of acid-base disturbance will have been mitigated. Problems may arise in interpreting data derived from blood taken from a patient who has achieved some degree of compensation. Unlike in the situation of a

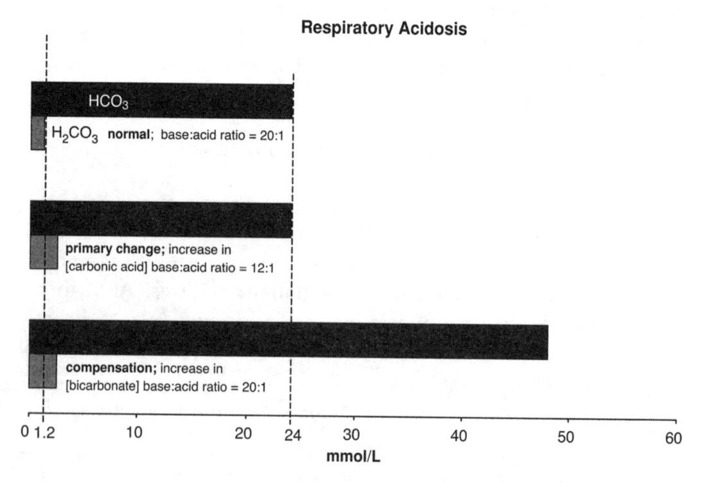

Figure 3.15

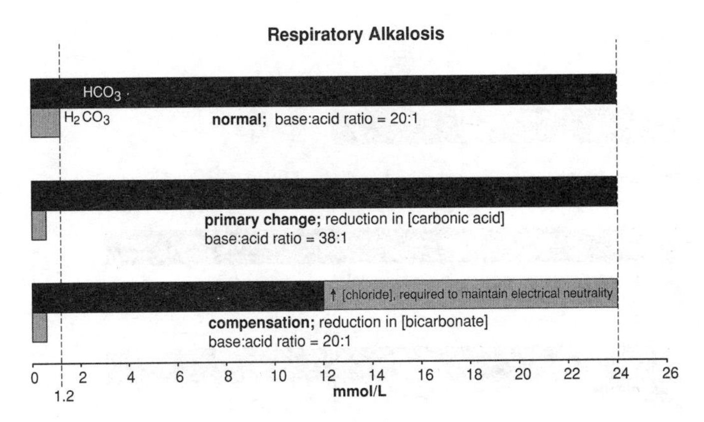

Figure 3.16

simple primary disturbance, during compensation the actual values of both PCO_2 and HCO_3^- will be abnormal and it can be difficult to decide the nature of the primary change and the compensatory response. The clinical history of the patient will usually provide some clues in such a situation. Figures 3.15 to 3.18 illustrate typical compensatory changes.

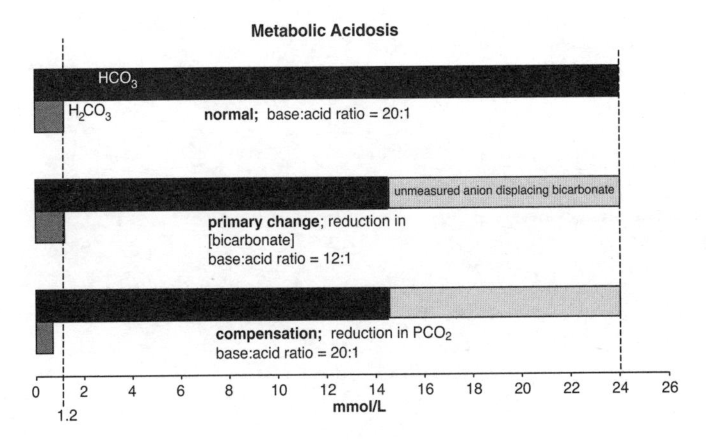

Figure 3.17

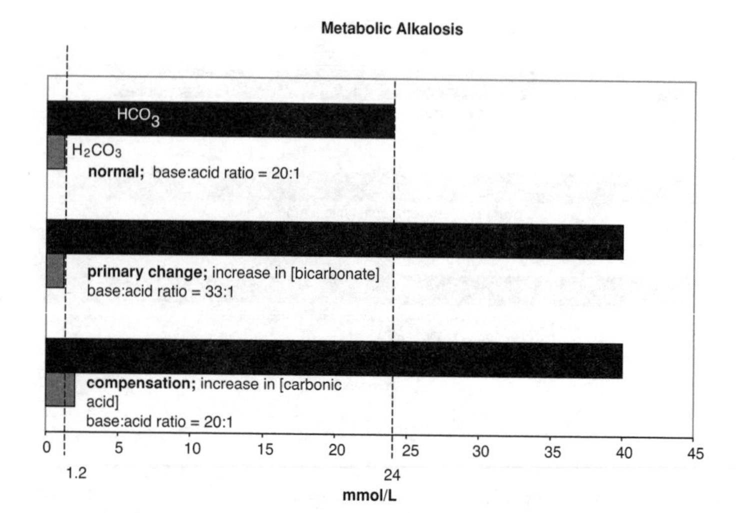
Metabolic Alkalosis
HCO3
H2CO3
normal; base:acid ratio = 20:1
primary change; increase in [bicarbonate]
base:acid ratio = 33:1
compensation; increase in [carbonic acid]
base:acid ratio = 20:1
0
5
10
15
20
25
30
35
40
45
1.2
24
mmol/L

Figure 3.18

SECTION 3.xvii

Check the data

Because of the mathematical interdependence, as defined by the Henderson-Hasselbalch equation, of the individual parameters used to assess acid-base status, it is easy to check the validity of results.

Given below are some equations that may be useful to cross-check results and confirm consistency.

1. $[H^+]$ nmol/L = antilog10^{9-pH}
2. $[H^+]\ \text{nmol/L} = \dfrac{PCO_2\text{kPa} \times 180}{[HCO_3^-]\ \text{mmol/L}}$
3. $[H^+]$ nmol/L = 80 – the decimal fraction of the pH.
   e.g. if pH = 7.26, 80 − 26 = 54 nmol/L
   Note: this method has limited accuracy but gives a value which should be ± 5 (approx.) of the $[H^+]$ when calculated using equations 1 or 2 above.
4. $[HCO_3^-]$ mmol/L = 0.225 × PCO_2 × $10^{pH-6.1}$
   (pH in the last term is the actual measured pH value)
5. Calculated PCO_2(in kPa) = (0.2 × HCO_3^- mmol/L) + 1.5
   Calculated PCO_2 should be ± 0.3 kPa of the measured PCO_2
   Note: All acid-base reference data assume the patient's body temperature is 37°C.
6. Calculate the anion gap if a metabolic disturbance is suspected.

$$AG = ([Na] + [K]) - ([Cl] + [HCO_3^-])$$

   Where the plasma albumin concentration is known or suspected to be abnormal a correction can be made:

$$AG_{corr} = \text{measured } AG + 0.25\ (41 - \text{measured [alb]})$$

Essential Fluid, Electrolyte and pH Homeostasis, First Edition. Gillian Cockerill and Stephen Reed.

SECTION 3.xviii

Self-assessment exercise 3.3

1. Case study 1: study the data given below:

	On admission	Ref.
pH	7.32	7.35–7.45
PCO_2	8.0 kPa	4.8–5.8 kPa
HCO_3^-	30 mmol/L	23–28 mmol/L

These results are consistent with:
(select ONE answer from those given)
(a) simple respiratory acidaemia
(b) simple respiratory alkalaemia
(c) simple non-respiratory acidaemia
(d) simple non-respiratory alkalaemia

2. Case study 2: study the data given below:

	On admission	Ref.
pH	7.48	7.35–7.45
PCO_2	3.5 kPa	4.8–5.8 kPa
HCO_3^-	19 mmol/L	23–28 mmol/L

These results are consistent with:
(select ONE answer)
(a) respiratory acidaemia
(b) respiratory alkalaemia
(c) non-respiratory acidaemia
(d) non-respiratory alkalaemia

Essential Fluid, Electrolyte and pH Homeostasis, First Edition. Gillian Cockerill and Stephen Reed.

3. Predict the type of mixed acid-base disturbance that might occur in someone who has suffered a cardiac arrest.

4. In order to compensate for a respiratory acidaemia, the body would:
   (select **ONE** answer from those given below)
   (a) Hyperventilate, to expel CO_2
   (b) Hypoventilate to retain CO_2
   (c) Retain more HCO_3^- in the kidney
   (d) Excrete more HCO_3^- via the kidneys

5. Using the Henderson-Hasselbalch equation, confirm validity of the data and then identify the type of acid-base disorder in each case (i) to (vi).

(i)
pH = 7.59
$[HCO_3^-]$ = 25 mmol/L
PCO_2 = 3.6 kPa

(ii)
pH = 7.14
$[HCO_3^-]$ = 12 mmol/L
PCO_2 = 4.8 kPa

(iii)
pH = 7.39
$[HCO_3^-]$ = 35 mmol/L
PCO_2 = 5.0 kPa

(iv)
pH = 7.42
$[HCO_3^-]$ = 27 mmol/L
PCO_2 = 5.8 kPa

(v)
pH = 7.57
$[HCO_3^-]$ = 14 mmol/L
PCO_2 = 2.1 kPa

(vi)

$$\begin{aligned} pH &= 7.12 \\ [HCO_3^-] &= 9\ mmol/L \\ PCO_2 &= 3.8\ kPa \end{aligned}$$

6. Which of the suggested causes is the most likely to be responsible for the data shown?

   Possible causes: diabetic coma
   acute renal failure
   chronic stimulation of the respiratory centre
   barbiturate overdose
   chronic respiratory failure

$$\begin{aligned} pH &= 7.20 \\ [HCO_3^-] &= 19\ mmol/L \\ PCO_2 &= 8.5\ kPa \end{aligned}$$

7. The 'ammonium chloride loading test' may be used to assess a subject's ability to acidify her/his urine. What is the physiological basis of this test?

SECTION 3.xix

Non-respiratory (metabolic) acidosis

Metabolic acidosis arises when the plasma bicarbonate concentration is reduced. In addition to the abnormalities in pH regulation, there are invariably associated changes that laboratory tests will reveal. For example, acidosis (of any origin) will *often* cause an increase in plasma potassium concentration (hyperkalaemia): see Table 3.2. The hyperkalaemia is partly due to the movement of K out of cells in exchange for protons which are buffered intracellularly. The hyperkalaemia and the reduced pH (if below approximately 7.10) both have deleterious effects on the contraction of the heart muscle. In addition to chemical buffering, the skeleton is used to mitigate acidosis by buffering protons with phosphate which is part of the hydroxyapatite bone mineral; this may lead to erosion and demineralisation of bone and thus to hypercalcaemia. Not only may the plasma *total* calcium concentration rise, but so too does the physiologically most active *ionised* fraction (see Section 2.viii) as the lower than normal pH causes dissociation of calcium bound to albumin;

$$\text{Alb} - \text{Ca} \xrightarrow{\text{Reduced pH}} \text{Alb}^- + \text{Ca}^{2+}$$

The anion gap (see Sections 1.x and 3.xiv) will increase if there is an accumulation of an 'unmeasured' anion such as lactate or hydroxybutyrate; the plasma chloride concentration will be normal. A normal anion gap is associated with hyperchloraemia; in effect, the chloride ion quantitatively displaces HCO_3^- from the plasma. Thus, the anion gap is a useful indicator of the origin of the acidosis.

Essential Fluid, Electrolyte and pH Homeostasis, First Edition. Gillian Cockerill and Stephen Reed.

Table 3.2 Some causes of metabolic acidosis

High anion gap	Normal anion gap	
	With hyperkalaemia	With hypokalaemia
Lactic acidosis due to hypoxia; Ketoacidosis, e.g. diabetes mellitus	Renal failure; Uraemic acidosis; Hypoaldosteronism	Diarrhoea; Types 1 and 2 renal tubular acidosis; Some diuretics

The compensatory mechanism for metabolic acidosis is hyperventilation; this response is driven by stimulation of blood pH sensors in the carotid arteries and central sensors detecting a fall in CSF pH.

(a) Diabetic ketoacidosis (DKA)

Diabetes mellitus arises due to either a lack of insulin (type 1) or due to tissue resistance to the stimulus of insulin (type 2). Insulin has many biochemical actions but the most well-known all relate to the utilisation of glucose as a fuel source. Glucose entry into most cells of the body is dependent upon insulin, and once inside the cell, insulin promotes glucose metabolism by influencing the activity of a number of key enzymes. When insulin is absent, glucose becomes, in effect, unavailable as a fuel source and so the tissues begin to use fat in greater quantities than they would normally.

The initial metabolic product of fatty acid metabolism is acetyl-CoA, which normally enters the Krebs tricarboxylic acid (TCA) cycle. However, the quantity of acetyl-CoA that can be used in this way is limited by the availability of oxaloacetate. When the number of acetyl-CoA molecules outstrips the oxaloacetate, any excess are converted into acetoacetate. In the liver, acetoacetate is then converted into hydroxybutyrate (see Figure 3.13). Acetoacetate and hydroxybutyrate are both acidic anions (acetoacetic acid and hydroxybutyric acid) and along with acetone (neutral) are known as ketone 'bodies'. The traditional model of acid-base imbalance explains the acidaemia that results as due to the accumulation of acidic metabolites. The Stewart model, on the other hand, views this as an increase in A_{TOT} and thus a decrease in SID (see Section 3.xii).

(b) Lactate acidosis

Lactate (lactic acid) is the end-product of anaerobic glycolysis. Lactate production is increased (i) during vigorous exercise, when glucose is the preferred fuel to power muscle contraction, or (ii) in situations when oxygen availability in cells is compromised (hypoxia; hypoxaemia PO_2 < 4 kPa). Reduced oxygen means that mitochondrial oxidation cannot function so NADH cannot be reoxidised; the TCA cycle cannot operate if NADH is not available. In short, normal ATP generation is seriously impaired. A small amount of ATP can be generated from glycolysis by the process known as substrate level phosphorylation, but this process requires reoxidation of the NADH produced by glyceraldehyde-3-phosphate dehydrogenase; reoxidation is achieved by the production of lactate by lactate dehydrogenase (see Figure 3.19).

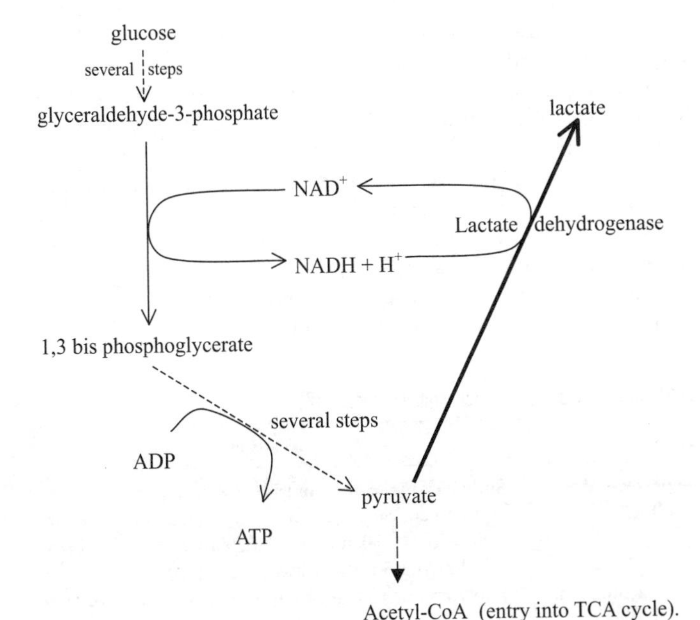

Figure 3.19 Overproduction of lactate during anaerobic metabolism. Lactate dehydrogenase re-oxidises NADH in order to maintain even modest ATP yield via glycolysis

The traditional interpretation is that accumulation of lactate anion displaces bicarbonate so its concentration is reduced to maintain electrical neutrality. On the other hand, the Stewart interpretation would be that lactate is a physiologically strong (but usually unmeasured) anion, so SID and pH both decrease and SIG increases (see page 265).

Lactate acidosis may occur when tissue hypoxia is *not* apparent, such as in diabetes mellitus or liver failure, due to the impaired ability to utilise normal amounts of lactate.

(c) Renal disease and dysfunction

Acidosis of renal failure is really due to acquired tubular dysfunction. As nephrons begin to fail, their ability to retain and regenerate bicarbonate and secrete protons into the tubular fluid in exchange for sodium diminishes. The principal cause of the acidosis is the impaired synthesis of ammonia from glutamine, so reducing the excretion of protons. The loss of sodium in the urine, plus the increased reabsorption of chloride to maintain electrical neutrality in the absence of bicarbonate, causes a reduction in the SID, but the anion gap may remain normal as chloride gain balances bicarbonate loss or may be increased if there is retention in the blood of sulphate and/or phosphate The loss of sodium, possibly exacerbated by some degree of aldosterone resistance in the distal tubule, contributes to the hyperkalaemia which is characteristic of renal failure when the glomerular filtration rate (GFR) falls significantly.

There are several causes of acquired tubular dysfunction, including acute tubular necrosis due to ischaemia following haemorrhage, and

Table 3.3 Classification of renal tubular acidoses (RTAs)

RTA	Features
Type 1 distal RTA	Subtypes due to autosomal recessive or autosomal dominant causes, hypokalaemia, plasma bicarbonate very low (<10 mmol/L), inability to acidify urine to below pH 5.5 following NH_4Cl load
Type 2 proximal RTA	Subtypes due to recessive or dominant causes, often part of a more generalised tubular defect, e.g. Fanconi syndrome, low/normal plasma bicarbonate concentration
Type 4	Hypoaldosteronism, hyperkalaemia

certain drugs (e.g. aminoglycosides and lithium) and toxins. Renal tubular acidosis (RTA) may be either genetic or acquired (e.g. toxins such as heavy metals or drugs or immunological disease). Three types (1, 2 and 4) are described; the original type 3 is now viewed as a variant of types 1 and 2 together. All types cause a systemic acidosis due largely to impaired ammonia generation (see Table 3.3).

(d) Drugs

(i) Alcohol intoxication

Alcohols are metabolised in the liver *via* oxidation reactions, firstly to aldehydes, then to the corresponding carboxylic acids:

$$\text{methanol} \longrightarrow \text{formaldehyde} \longrightarrow \text{formate (formic acid)}$$
$$\text{ethylene glycol} \longrightarrow \text{glycoaldehyde} \longrightarrow \text{glycolate} \longrightarrow \text{oxalate (oxalic acid)}$$

Ethanol is metabolised in a similar fashion to acetaldehyde and acetate, but acetate may be further metabolised as acetyl-CoA and finally to carbon dioxide.

The traditional interpretation is overproduction of organic acid and accumulation of anions displacing HCO_3^-. The Stewart interpretation would be as for DKA: loss of sodium as a counter cation for the organic anion causes a reduction in SID and/or an increase in total weak acids, A_{TOT}.

(ii) Aspirin (acetylsalicylic acid)

Aspirin is metabolised to salicylate and an overdose of the drug leads to a *mixed* acid-base disturbance: metabolic acidosis coincident with a respiratory alkalosis. The acidosis is usually more significant in children than in adults. The salicylate anion probably contributes relatively little to the metabolic disturbance, its main effect being via an increase in lactate and ketoacids.

(e) Diarrhoea

The secretions into the gut below the stomach are alkaline; for every bicarbonate ion that is secreted into the intestine, a proton is added to the bloodstream. Much of the secreted bicarbonate-rich fluid would normally be reabsorbed to help to balance the proton load, but the reduced transit time of gut contents that characterises diarrhoea does not allow this to happen. Thus, the blood is acidified and bicarbonate is lost.

The situation may be complicated by the response to the loss of fluid. Reduction in ECF volume initiates aldosterone secretion, stimulating reclamation of sodium in the nephron but also loss of potassium, resulting in hypokalaemia.

SECTION 3.xx

Metabolic acidosis: detailed case studies

1. A 65-year-old male in acute renal failure and with a history of alcohol abuse.

	On admission	Ref.
Whole blood		
pH	7.20	7.35–7.45
$[H^+]$	63 nmol/L	35–45 nmol/L
PCO_2	4.4 kPa	4.8–5.8 kPa
HCO_3^-	15 mmol/L	23–28 mmol/L
BE	−10 mmol/L	+2 to −2 mmol/L
Plasma		
Na^+	137 mmol/L	135–145 mmol/L
K^+	6.0 mmol/L	3.5–4.5 mmol/L
Cl^-	112 mmol/L	97–107 mmol/L
Urea	31.2 mmol/L	3.0–7.0 mmol/L

Interpretation: Clearly this is an acidaemia caused by the acute renal failure (the acidosis). The low BE and the relatively normal PCO_2 indicate this is a metabolic problem. A slightly lower PCO_2 might have been expected assuming the respiratory system is fully functional. The 'traditional' interpretation is an acidosis due to failure in the production of ammonium from glutamine in the damaged nephron cells, and thus inability to reabsorb or generate bicarbonate and secrete protons. The implicit liver damage due to alcohol abuse may affect lactate uptake and utilisation.

The apparent SID is $\{137 + 6.0\} - 112 = 31$ mmol/L, which is 9 mmol/L lower than the norm of 40 mmol/L, a value which agrees well with the BE of −10 so we can be sure there is no

Essential Fluid, Electrolyte and pH Homeostasis, First Edition. Gillian Cockerill and Stephen Reed.

respiratory factor contributing to the imbalance. The Stewart interpretation of the underlying mechanism of the acidaemia is the excess chloride (hyperchloraemic acidosis with low SID) and possibly some weak (phosphate) and/or strong (sulphate or lactate) acid anions (increased SIG, see page 265) which are not being excreted by the impaired kidney.

The same clinical interpretation is reached whether using the traditional or Stewart approach. The difference between them is in the explanation of the underlying mechanism; the traditional model explains this as a primary bicarbonate deficiency, whilst Stewart sees it as chloride excess. The usefulness of the BE and the SID together are apparent in this example.

2. A 55-year-old homeless female admitted unconscious. Medical staff had reason to believe that the patient had taken a 'cocktail' of liquids, including methanol and unknown drugs in a suicide attempt. Arterial blood was drawn for analysis.

	On admission	Ref.
Whole blood		
pH	7.21	7.35–7.45
$[H^+]$	62 nmol/L	35–45 nmol/L
PCO_2	6.8 kPa	4.8–5.8 kPa
PO_2	8.2 kPa	11–13 kPa
HCO_3^-	14 mmol/L	23–28 mmol/L
BE	−11 mmol/L	+2 to −2 mmol/L
Plasma		
Na^+	138 mmol/L	135–145 mmol/L
K^+	5.0 mmol/L	3.5–4.5 mmol/L
Cl^-	115 mmol/L	97–107 mmol/L
Urea	6.9 mmol/L	3.0–7.0 mmol/L
Osmol gap	35 mmol/L	<10 mmol/L

Interpretation: Low values for pH and actual bicarbonate indicate an acidosis, and the negative BE suggests there is a metabolic component to the acidosis. However, the PCO_2 is high, indicating a respiratory factor is also present. In the presence of a metabolic acidaemia, respiratory compensation would cause hyperventilation and so reduce the PCO_2. Clearly this has not happened, so there is some defect in the respiratory chain preventing hyperventilation.

Given the circumstances, this could be drug-induced suppression of the respiratory centre. The SID is $\{148 + 5\} - 115 = 38$ mmol/L; $40-38 = 2$ which, compared with the BE of -11, suggests that chloride is *not* the only strong anion present: possibly lactate is present as a result of the hypoxia (implicit from the hypoxaemia, low PO_2), or formate if methanol ingestion was significant.

This seems to be a mixed acidosis with a toxin-induced metabolic component and the high PCO_2 contributing to a respiratory acidosis. The very high osmolality gap (difference between the measured osmolality and the osmolality calculated from electrolytes, urea and glucose concentrations: see Section 1.xiii) is evidence for the presence in the plasma of an unmeasured 'osmolyte', probably in this case methanol or ethanol.

3. Terry is 32 years old and works for a merchant bank in the City. Recently he had been working more hours than usual. A known diabetic, he is normally well-controlled on insulin injections. His girlfriend returns home one day to find Terry unconscious on the floor of their flat. She calls an ambulance and on arrival at hospital, arterial blood was taken.

	On admission	Ref.
Whole blood		
pH	7.15	7.35–7.45
$[H^+]$	71 nmol/L	35–45 nmol/L
PCO_2	3.4 kPa	4.8–5.8 kPa
PO_2	11.2 kPa	11–13 kPa
HCO_3^-	10 mmol/L	23–28 mmol/L
BE	−14 mmol/L	+2 to −2 mmol/L
Plasma		
Na^+	141 mmol/L	135–145 mmol/L
K^+	5.5 mmol/L	3.5–4.5 mmol/L
Cl^-	100 mmol/L	97–107 mmol/L
Urea	6.9 mmol/L	3.0–7.0 mmol/L
Glucose	31.2 mmol/L	
Osmol gap	15 mmol/L	<10 mmol/L
Anion gap	33 mmol/L	10–15 mmol/L

Interpretation: Diabetic ketoacidosis. Terry had apparently forgotten to take his insulin, which with the stress of work had thrown his diabetic control out of balance. The elevated anion gap can be accounted for by accumulation of metabolic acids.

4. Mrs S had been resident in a care home for 18 months. She had had recurrent urinary tract infections over recent weeks. Nursing staff found her in a semi-comatose state and noted she had very cold hands; an ambulance was called. Additionally, Mrs S had a raised temperature, a heart rate of 140 bpm and a blood pressure of 80/42 mmHg.

	On admission	Ref.
Whole blood		
pH	7.19	7.35–7.45
$[H^+]$	71 nmol/L	35–45 nmol/L
PCO_2	4.2 kPa	4.8–5.8 kPa
PO_2	11.6 kPa	11–13 kPa
HCO_3^-	12 mmol/L	23–28 mmol/L
BE	−11 mmol/L	+2 to −2 mmol/L
Plasma		
Na^+	134 mmol/L	135–145 mmol/L
K^+	5.1 mmol/L	3.5–4.5 mmol/L
Cl^-	102 mmol/L	97–107 mmol/L
Urea	7.9 mmol/L	3.0–7.0 mmol/L
Glucose	6.4 mmol/L	
Osmol gap	12 mmol/L	<10 mmol/L
Anion gap	22 mmol/L	10–15 mmol/L
Lactate	5.5 mmol/L	<2.5 mmol/L

Interpretation: A provisional diagnosis of septic shock was made by medical staff in A&E and this was later confirmed by bacteriological results. Despite the normal PO_2, Mrs S showed signs of tissue hypoxia due to poor peripheral circulation, which explains the raised plasma lactate concentration, and therefore the increase in the anion gap.

SECTION 3.xxi

Non-respiratory (metabolic) alkalosis: overview

An elevation of plasma bicarbonate concentration is always an indication of metabolic alkalaemia, unless it is the result of compensation for a respiratory acidosis. The nephron normally reabsorbs bicarbonate very efficiently up to the threshold of approximately 27 mmol/L; beyond that value, bicarbonaturia occurs. However, disorders that result in a metabolic alkalaemia do so by raising the renal threshold so bicarbonate excretion does not occur. It is the elevation in the renal threshold that *maintains* rather than *initiates* the alkalaemia, i.e. high plasma bicarbonate concentration. Simple metabolic alkalosis is relatively uncommon, there being only a few pathological circumstances that initiate a rise in bicarbonate concentration.

Initiating causes of metabolic alkalaemia include:

1. Changes to the renal handling of protons as a consequence of changes in the renal handling of potassium: e.g. diuretics, mineralocorticoid excess (primary hyperaldosteronism, Conn's syndrome) or Cushing's syndrome (cortisol has a weak mineralocorticoid effect);
2. Proton loss via the gut, e.g. loss of gastric contents, diarrhoea fluid which is rich in chloride;
3. Increases in exogenous alkali:
   (a) ingestion of alkali (e.g. antacids for indigestion) or intravenous therapy for acidosis;

Essential Fluid, Electrolyte and pH Homeostasis, First Edition. Gillian Cockerill and Stephen Reed.

or

(b) ingestion of substances which are metabolised into bicarbonate, e.g. many weak organic anions such as citrate, lactate and acetate. Surprisingly perhaps, the ingestion of weak acids does not constitute an acidosis but an alkalosis;

4. Miscellaneous: reduction of ECF volume ('contraction alkalosis').

Maintenance of the alkalaemia, that is, increased bicarbonate reabsorption in the nephron invariably involves hyperaldosteronism, hypovolaemia and hypokalaemia, so the situation can become self-sustaining.

Mechanisms of maintenance of alkalaemia

Physiological compensation for metabolic alkalaemia is hypoventilation. Central and peripheral pH sensors slow the respiratory rate. This, however, is somewhat self-limiting because, assuming respiratory control is fully functional, a rise in PCO_2 to around 7.5 kPa (or fall in PO_2) will trigger hyperventilation (see Figure 3.6, Section 3.vii).

Hypokalaemia causes K^+ to leak from cells in exchange for protons, resulting in an intracellular acidosis, which induces the renal tubular cells to reabsorb more bicarbonate from the tubular fluid. Additionally, hypokalaemia is believed to stimulate the action of an ATP'ase H^+/K^+ exchange mechanism in the distal tubules; K^+ is reabsorbed at the expense of proton loss. The effect of both mechanisms is to raise the renal threshold for bicarbonate.

Hypovolaemia, which may be due to fluid loss through vomiting or 'pooling' of plasma fluid in ascites or oedema, triggers a predictable homeostatic response. There will be increased production of angiotensin II, aldosterone and adrenalin, all of which stimulate Na^+ reabsorption in exchange for H^+ in the proximal nephron.

Hyperaldosteronism, possibly as a response to volume contraction as discussed above, promotes Na^+ reabsorption and K^+ excretion, so exacerbating the hypokalaemia and proton excretion.

SECTION 3.xxii

Non-respiratory (metabolic) alkalosis: causes

(a) Vomiting

Vomiting causes loss of gastric fluid containing H^+, Na^+ and K^+ (see Table 1.6 in Section 1.ix). The loss of K^+ in this way is relatively more important than the loss of Na^+ because the concentrations of the two ions present in gastric juice and the loss of protons causes the alkalaemia. The Stewart approach would classify this condition as a loss of chloride and so an increase in the SID.

Thus, the initiating factor (the alkalosis) is the loss of acid and the alkalaemia is maintained by hypokalaemia (renal losses) stimulated by aldosterone, itself induced by hypovolaemia. A similar biochemical picture may be seen in patients undergoing gastric lavage and aspiration, perhaps following a drug overdose or toxin ingestion.

The appropriate treatment for this condition is the infusion of saline.

(b) Diuretics

Diuretic drugs are widely used to reduce blood volume in conditions such as congestive cardiac failure or liver cirrhosis. Those particular drugs known as thiazides (e.g. chlorthiazide) and loop diuretics (e.g. ethacrynic acid and furosemide) inhibit NaCl reabsorption in the proximal tubule or the loop of Henle, but cause kaluria (K^+ loss in urine) and therefore hypokalaemia. Additionally, there is an induced

Essential Fluid, Electrolyte and pH Homeostasis, First Edition. Gillian Cockerill and Stephen Reed.

hypochloraemia and, of course, ECF contraction effects which both contribute to the alkalaemia.

In contrast, spironolactone, an aldosterone antagonist, and other so-called potassium-sparing diuretics such as amiloride cause *hyper*kalaemia. Careful laboratory-based monitoring of electrolyte concentrations is required for patients on long-term diuretic therapy.

(c) Mineralocorticoid excess

Hyperaldosteronism (Conn's Syndrome) and chemicals that mimic the effects of aldosterone such as glycerrhizic acid (an active component of liquorice) all cause potassium loss through the kidney, and thus a metabolic alkalaemia by the mechanisms discussed above.

Case studies

1. Pyloric stenosis is a condition in which the valve (pylorus) between the stomach and the duodenum fails to open properly. Gastric contents cannot pass into the small intestine so reflex vomiting occurs. A 9-month-old female infant was admitted following two days of vomiting. Results on admission and 12 hours later following saline infusion were as shown:

	On admission	After 12 hours	Ref.
Whole blood			
pH	7.56	7.48	7.35–7.45
PCO_2	6.5 kPa	5.8 kPa	4.8–5.8 kPa
Actual HCO_3^-	44 mmol/L	32 mmol/L	22–28 mmol/L
Plasma			
Na^+	130 mmol/L	138 mmol/L	135–145 mmol/L
K^+	2.4 mmol/L	3.0 mmol/L	3.5–4.5 mmol/L
HCO_3^-	43 mmol/L	33 mmol/L	22–28 mmol/L
Cl^-	70 mmol/L	95 mmol/L	97–107 mmol/L

Interpretation: The traditional interpretation of this is loss of gastric HCl leading to a net excess of bicarbonate. Recall that for every proton secreted by the parietal cells into the gastric lumen, a HCO_3^- ion is added to the plasma. The elevated PCO_2 on admission is respiratory compensation which may have resulted in some degree of hypoxia. The loss of sodium and water (hypovolaemia) trigger an aldosterone response which causes increased reabsorption of NaCl in the nephron and induces hypokalaemia.

2. Mr AH, a 55-year-old, consulted his GP due to slowly progressive abdominal swelling and increasing discomfort. Clinically, he showed some peripheral oedema. His previous medical history included a severe flu-like illness soon after an overseas trip. The GP prescribed diuretics to reduce the oedema. A subsequent blood test revealed:

	On admission	Ref.
Plasma		
Na^+	144 mmol/L	135–145 mmol/L
K^+	3.0 mmol/L	3.5–4.5 mmol/L
HCO_3^-	36 mmol/L	23–28 mmol/L

The oedema fluid that collected in the plural space contained all the ions found in plasma. When the diuretics began to be effective, the patient excreted a bicarbonate-poor fluid and the HCO_3^- which had been within the oedema fluid was returned to the plasma. The situation is often referred to as a 'contraction alkalosis', i.e. the contraction in ECF volume has resulted in redistribution of electrolytes and a rise in bicarbonate concentration. An additional factor may be that certain diuretics cause hypokalaemia, which itself leads to alkalaemia (Section 2.xxv). A full blood gas and acid-base assessment is not warranted in such a case.

SECTION 3.xxiii

Self-assessment exercise 3.4

1. What is the physiological reasoning for the use of saline infusion to treat loss of gastric juice?
2. What effect would the infusion of a litre of 'normal' physiological saline have on the SID? *(HINT: physiological saline contains only NaCl at a concentration of approximately 150 mmol/L.)*
3. How would the Stewart model explain the acid-base imbalance in pyloric stenosis?

Check your answers before progressing.

Essential Fluid, Electrolyte and pH Homeostasis, First Edition. Gillian Cockerill and Stephen Reed.

SECTION 3.xxiv

Respiratory disorders: overview

The physical act of breathing requires normal functioning of three physiological systems: (i) the respiratory centre in the brain (central control), (ii) neural impulse transmission from the respiratory centre via the spinal cord and phrenic nerve, to (iii) the intercostal muscles (rib cage) and muscles of the diaphragm whose contraction expands the thorax to allow air to enter the lungs; relaxation of the same muscles forces air out of the lungs. Defects in any part of the 'chain' will affect ventilation. In addition, damage to or abnormality of the lungs themselves, sometimes called intrinsic pulmonary disease, will affect gaseous exchange across the alveolar/capillary barrier.

Respiratory disorders are characterised by abnormalities in the PCO_2. Higher then usual production of carbon dioxide occurs during physical exertion as more substrates are oxidised within cells to generate energy (ATP) needed to meet the demands. Overproduction of CO_2 does not occur pathologically, so the only reason for hypercapnia (i.e. an elevated PCO_2) is reduced excretion due to (a) impaired diffusion at the alveoli, (b) reduced respiratory rate, or (c) reduced tidal volume, which is the volume of air that enters and leaves the lungs each minute.

$$\text{Pulmonary ventilation} \propto \text{number of breaths per minute} \times \text{volume of each breath} (= \text{tidal volume}) \text{ typically, } 15 \times 500\,\text{mL} = 7.5\,\text{litres/minute.}$$

Thus, shallow and/or slow breathing reduces ventilation and so increases PCO_2. Conversely, hypocapnia, a low PCO_2, is due to increased pulmonary ventilation, usually as a result of more rapid rather than deeper breathing.

Essential Fluid, Electrolyte and pH Homeostasis, First Edition. Gillian Cockerill and Stephen Reed.

As a rule, acute respiratory distress causes an initial alkalosis as physiological feedback mechanisms try to increase ventilation, whereas chronic conditions in which respiratory muscles are weakened or fatigued result in an acidosis.

Compensation for respiratory disorders is via the kidney: changes in proton secretion and bicarbonate reabsorption/generation. These mechanisms may take up to 48 hours to become effective.

Specific causes of respiratory disorders are shown in Table 3.4.

Table 3.4 Some common causes of respiratory disorders

Origin	Respiratory acidosis	Respiratory alkalosis
Central control	Depression of respiratory centre due to e.g. drugs (barbiturate, morphine, hypnotics, narcotics, anaesthetics) CNS infections CNS trauma (head injury) or infarction	Hypoxia, hypoxaemia Emotional and psychological, e.g. anxiety, hysteria Stimulant drugs, e.g. salicylate (mixed) Pregnancy (progesterone) CNS infection, tumour Systemic fever
Neuromuscular	Polio Multiple sclerosis Myopathy Hypokalaemia Muscle relaxants	None
Thoracic disease	Damage to chest wall Adult or infant respiratory distress syndrome	None
Pulmonary	Airway obstruction Asthma Bronchitis Emphysema Severe oedema Foreign body or other blockage	Embolism Asthma Pneumonia Oedema
Cardiac	Cardiac arrest (mixed disturbance, page 278)	Congestive heart failure

SECTION 3.xxv

Physiological consequences of respiratory disorders

Compensation for respiratory disorders is via the kidney. Hypercapnia (primary respiratory acidosis) leads to increased intracellular PCO_2, so carbonic anhydrase in the tubular cells produces more HCO_3^- (added to the plasma) and increased ammonia production. These two mechanisms allow increased proton secretion into the lumen. Also, there is reduced NaCl reabsorption in the proximal tubule and a tendency to kaliuria in chronic respiratory acidosis, whereas acute respiratory acidosis is associated with hyperkalaemia due in part to decreased potassium excretion in the nephron. Potassium and phosphate released from cells (both contribute to buffering of the excess protons) usually lead to hyperkalaemia and hyperphosphataemia.

Hypercapnia has effects in major organs such as the heart and the brain. Acidaemia suppresses the contractility of cardiac myocytes (muscle cells) and hypercapnia-induced vasodilation in the CNS causes symptoms ranging from headache to drowsiness to coma.

Respiratory alkalosis (hypocapnia) leads to reduced HCO_3^- reabsorption, bicarbonaturia and diminished net acid excretion. In the acute phase of hypocapnia there is a slight fall in plasma bicarbonate concentration as the reduced PCO_2 means that the 'key equation'

$$CO_2 + H_2O \longleftrightarrow H_2CO_3 \longleftrightarrow H^+ + HCO_3^-$$

actually moves from right to left. Presumably, protons are released from non-bicarbonate buffers such as proteins or phosphate to

Essential Fluid, Electrolyte and pH Homeostasis, First Edition. Gillian Cockerill and Stephen Reed.

replenish those consumed. Increased glucose metabolism *via* glycolysis allows accumulation of lactate with a slight rise in the anion gap.

Potassium and phosphate both enter cells bringing about hypokalaemia and hypophosphataemia respectively. Effects on the brain and heart are the opposite of those described above; cerebral vasoconstriction causes light-headedness, or more severely seizures, and cardiac output may fall.

SECTION 3.xxvi

Respiratory disorders: case studies

(a) Respiratory acidosis

(1) A 23-year-old woman was admitted to Accident & Emergency having been knocked down in the street by a motorcycle. The impact was not great, but as she fell, the patient's head made a heavy contact with the kerb and she was knocked unconscious. The ambulance crew who attended noticed an abnormally low respiratory rate.

	On admission	Ref.
Arterial whole blood		
pH	7.07	7.35–7.45
PCO_2	8.6 kPa	4.8–5.8 kPa
HCO_3^-	18 mmol/L	23–28 mmol/L

These data, along with the clinical history, indicate an obvious respiratory acidaemia. The blow to the back of the head has damaged the respiratory centre which is no longer sending nerve impulses to the respiratory muscles of the thorax to induce contraction. Pulmonary gas exchange is impaired and CO_2 is being retained. The reduction in the bicarbonate reflects the low pH.

(2) Mike was a known asthmatic and was well used to dealing with mild acute attacks by the use of his aspirator containing a bronchodilator. Recently he has contracted a severe chest infection which has caused breathing difficulties and a degree of discomfort. When an acute asthma attack struck he was unable to relieve

Essential Fluid, Electrolyte and pH Homeostasis, First Edition. Gillian Cockerill and Stephen Reed.

the symptoms with the aspirator and he began to panic as his breathing became progressively more laboured.

	On admission	Ref.
Arterial whole blood		
pH	7.19	7.35–7.45
PCO_2	7.7 kPa	4.8–5.8 kPa
PO_2	9.8 kPa	11–14 kPa
HCO_3^-	21 mmol/L	23–28 mmol/L

This is a case of acute on chronic airways disease. The bronchitis had caused some degree of airway narrowing and the acute attack resulted in a more severe constriction, impairing gas exchange. Had the condition progressed further it is possible that the rising PCO_2 and falling PO_2 would have caused stimulation of the respiratory centre and hyperventilation. Chemoreceptors that 'detect' pH and/or PCO_2 respond in an exaggerated way when there is hypercapnia with simultaneous hypoxaemia.

(3) Abi was a 20-year-old student who went to an end-of-term party with her housemates. She not only consumed far more alcohol than she could manage, but also took some drugs, believed to be 'sleeping pills'. She was found by a fellow partygoer, unconscious and with very shallow respiration.

	On admission	Ref.
Arterial whole blood		
pH	7.20	7.35–7.45
PCO_2	6.8 kPa	4.8–5.8 kPa
PO_2	13.1 kPa	11–14 kPa
HCO_3^-	20 mmol/L	23–28 mmol/L

Abi has a respiratory acidaemia arising from a reduction in the central neural drive. The combination of the drugs, which may have been barbiturates, and alcohol has depressed the respiratory centre and thus the respiratory rate.

(b) Respiratory alkalosis

(1) A 40-year-old male was admitted following a failed suicide attempt. He had been found in his fume-filled car which had a piece of hose leading from the exhaust pipe to the interior. The man was clearly hyperventilating.

	On admission	Ref.
Arterial whole blood		
pH	7.71	7.35–7.45
PCO_2	2.6 kPa	4.8–5.8 kPa
PO_2	10.2 kPa	11–14 kPa
HCO_3^-	24 mmol/L	23–28 mmol/L

Unsurprisingly, the blood sample had a very high concentration of carbon monoxide. The hyperventilation in this case was due to tissue hypoxia even though there is only a slight hypoxaemia. Binding of carbon monoxide to haemoglobin has reduced oxygen delivery to the tissues, and in an attempt to correct this situation the respiratory centre is stimulated. The PCO_2 is reduced by the hyperventilation causing a rise in blood pH.

(2) Charles, a 58-year-old accountant with a large company experienced, vague feelings of myalgia whilst travelling home from the office one Friday evening in early January. His discomfort and general weakness increased over the weekend and by Sunday evening he developed periods of uncontrollable shivering despite having a high temperature. Two weeks later his symptoms had not improved, and indeed his breathing had become shallow and rapid and the weakness more severe. The GP was called and her assessment was that Charles' respiratory symptoms were sufficiently severe for him to be transferred to hospital as an emergency.

	On admission	Ref.
Arterial whole blood		
pH	7.54	7.35–7.45
PCO_2	3.4 kPa	4.8–5.8 kPa
PO_2	5.3 kPa	11–14 kPa
SO_2	79%	>95%

Electrolyte results and the anion gap were all normal. Charles was treated with 100% oxygen and transferred to the Intensive Care Unit, given respiratory support and treated with antibiotics for a secondary bilateral bacterial pneumonia. One week later and now breathing 35% oxygen, arterial blood analysis gave the following results:

	After 1 week	Ref.
Arterial whole blood		
pH	7.47	7.35–7.45
PCO_2	3.8 kPa	4.8–5.8 kPa
PO_2	10.1 kPa	11–14 kPa
SO_2	96%	>95%

Again electrolytes were mostly normal (only the base excess was slightly out of range at −4 mmol/L) but there was a mild anaemia:

	After 1 week	Ref.
Arterial whole blood		
Haemoglobin	10.6 g/dL	12.5–16 g/dL
Haematocrit	34%	35–50%

Mean red cell haemoglobin concentration and mean red cell volume were normal but the red cell count was low.

Interpretation: This is a clear case of respiratory alkalaemia. The physiological problem within the lungs lies not in the detection of, or response to, inadequate oxygenation but is pulmonary. The hyperventilation is this case was probably due to a combination of fever arising from the infection and inappropriate stimulation of stretch receptors within the lung by accumulation of phlegm and purulent debris. Such stimulation triggers the respiratory centre to increase neural output resulting in hyperventilation. In addition, the lung infection caused poor oxygen transfer into the blood; note that on admission the arterial PO_2 is comparable to that we would expect to find in venous blood and the SO_2 is very low. This oxygen deficit prompted hyperventilation in a vain attempt to increase oxygen intake, but the consequence of hyperventilation is hypocapnia (low PCO_2) and thus an alkalaemia.

As the infection began to resolve, oxygen support was gradually reduced but SO_2 and PO_2 were maintained at reasonable values despite the persistently low PCO_2 and slightly high pH (as shown by the second set of data). The base excess indicates that some degree of renal compensation was occurring and the haematological data suggest a normocytic anaemia. Just four days later, Charles was able to maintain adequate oxygenation whilst breathing ordinary air, and after another week on ICU he was transferred to a general ward.

SECTION 3.xxvii

Summary of Part 3

- pH homeostasis is one of the most sensitive regulatory mechanisms animals possess.
- Acid-base balance requires the subtle interplay of a number of physiological and biochemical mechanisms.
- The production of carbon dioxide (which is hydrated to form carbonic acid), organic and inorganic acids derived from cellular metabolism each add significant amounts of protons to our body fluids.
- Initially, the protons are buffered by proteins and the bicarbonate/carbonic acid system.
- The kidneys, lungs and respiratory system, liver and gut all play significant roles in acid-base homeostasis.
- A large proportion of the 150 mol of protons produced each day are metabolically recycled or used in the kidney to reabsorb and regenerate bicarbonate. The kidney excretes approximately 70 mmol (70 mEq) of protons each day.
- The kidneys acidify the tubular fluid and excrete protons buffered as $H_2PO_4^-$ and NH_4^+. The maximum limit of urinary acidification the kidney can achieve corresponds to pH 4.5.
- Adequate supplies of sodium to the tubular cells of the nephron are essential for normal proton secretion and bicarbonate reabsorption.
- The 'traditional' approach to understanding pH homeostasis relies upon the Henderson-Hasselbalch equation. This works for many practical purposes. This buffer system only operates well *in vivo* because we have control over the bicarbonate component via the kidney and the carbonic acid component via respiration.
- The 'modern' approach is based on rigorous mathematical and chemical reasoning and views the changes in sodium, potassium and chloride (SID), PCO_2 and A_{TOT} as the driving forces in acid-base homeostasis via their influence on the dissociation of water.

Essential Fluid, Electrolyte and pH Homeostasis, First Edition. Gillian Cockerill and Stephen Reed.

- Acidaemia and alkalaemia are symptoms of underlying acidosis or alkalosis respectively.
- There are five classes of acid-base disturbance, and compensation is the process the body uses in an attempt to reverse the primary upset, which in the traditional model is based on re-establishing the 20:1 base-to-acid ratio as defined by the Henderson-Hasselbalch equation.
- The traditional and Stewart approaches differ mainly in how they interpret metabolic disturbance and there is no disagreement over interpretation of respiratory conditions.
- Interpretation of acute conditions is usually straightforward but in chronic conditions when some degree of compensation has occurred, data may not be so easy to understand.

Answers to Part 3 self-assessment exercises

Self-assessment exercise 3.1

1. (a) Confirm that the reference range for arterial blood $[H^+]$ is 35–45 nmol/L.

   Given that $[H^+]$ nmol/l = antilog $10^{9\text{-pH}}$
   The stated reference range for arterial blood pH is 7.35–7.45
   Substituting when pH = 7.35
   $[H^+]$ nmol/l = antilog $10^{9-7.35}$
   antilog $(9 - 7.35) = 44.7$ nmol/L(~ 45 nmol/L)
   Substituting when pH = 7.45
   $[H^+]$ nmol/l = antilog $10^{9-7.45}$
   antilog $(9 - 7.45) = 35.4$ nmol/L

   (b) What is the $[H^+]$ gradient across the plasma membranes of cells in contact with interstitial fluid (assume this to be identical in composition to venous blood)?

   ICF pH is 6.85 = 141.2 nmol/L, i.e. antilog $10^{9-6.85}$
   Venous blood/ISF pH is 7.35 = 44.7 nmol/L, i.e. antilog $10^{9-7.35}$
   Therefore the gradient = 3.16:1 $(141.2 \div 44.7)$

Essential Fluid, Electrolyte and pH Homeostasis, First Edition. Gillian Cockerill and Stephen Reed.

2. Verify that $pH = pK_a$ when [base] = [conjugate acid].

$$pH = pK_a + \log \left\{ \frac{[\text{base}]}{[\text{conj. weak acid}]} \right\}$$

if [base] = [acid], the ratio = 1

$\log 1 = 0$

$\therefore pH = pK_a + 0$ so $pH = pK_a$

3. Given pK_a for lactic acid = 3.8, what is the lactate:lactic acid ratio inside a cell under normal biochemical conditions?

$$pH = pK_a + \log \left\{ \frac{[\text{base}]}{[\text{conj. weak acid}]} \right\}$$

normal *cellular* pH = 6.85

substituting:

$$6.85 = 3.8 + \log \left\{ \frac{[\text{lactate}]}{[\text{lactic acid}]} \right\}$$

$6.85 - 3.8 = 3.05 = \log$ [lactate]/[lactic acid]

antilog 3.05 = 1122

the lactate:lactic acid ratio $\sim 1120 : 1$

4. The normal range for arterial blood pH is 7.35 – 7.45 but individuals are able to survive pH values as low as 7.10 or as high as 7.70.
   (a) What is the $[H^+]$ in a blood sample whose pH is 7.10?

$$[H^+] = 80 \text{ nmol/L}$$

   (b) What is the $[H^+]$ in a blood sample whose pH is 7.70?

$$[H^+] = 20 \text{ nmol/L}$$

These values can be calculated using several ways, but if we recall that a change in pH of 0.3 units is equivalent to a 2-fold change in $[H^+]$, calculation is not necessary:

$7.40 - 7.10 = 0.3$, therefore the $[H^+]$ doubles to 80 nmol/L
$7.40 - 7.70 = -0.3$, therefore the $[H^+]$ halves to 20 nmol/L

5. Given that pK_a for H_2CO_3 *in vivo* = 6.10, calculate the base to acid *ratio* at:
   (a) the upper limit of normal for arterial pH;
   (b) the lower limit of normal for arterial pH.

$$pH = pK_a + \log \left\{ \frac{[\text{base}]}{[\text{conj. weak acid}]} \right\}$$

Substituting at the upper limit of normal blood pH

$$7.45 = 6.1 + \log [HCO_3^-]/[H_2CO_3]$$
$$7.45 - 6.1 = 1.35 = \log [HCO_3^-]/[H_2CO_3]$$
$$\text{antilog } 1.35 = [HCO_3^-]/[H_2CO_3]$$
$$\therefore \text{ ratio} = 22.4$$

Substituting at the lower limit of normal blood pH

$$7.35 = 6.1 + \log [HCO_3^-]/[H_2CO_3]$$
$$7.35 - 6.1 = 1.25 = \log [HCO_3^-]/[H_2CO_3]$$
$$\text{antilog} 1.25 = [HCO_3^-]/[H_2CO_3]$$
$$\therefore \text{ ratio} = 17.8$$

6. An astute junior doctor notices that the $[H^+]$ of an arterial blood sample is reported as 65 nmol/L but the pH is given as 7.25. Is she justified in asking for the result to be confirmed?

   Yes, she is justified. A pH of 7.25 corresponds to a $[H^+]$ of 56 nmol/L and a 65 nmol/L corresponds to a pH of 7.19. It appears that a transcription error has been made in reporting the data on the patient.
7. Urine contains approximately the same amounts of phosphate and sulphate, yet only phosphate is able to act as an effective buffer. Suggest a reason for this.

   With the exception of bicarbonate, an anion can only act effectively as a buffer if its conjugate acid has a pK_a within ± 1 pH units of the 'target' pH. Given urine pH is typically in the range 5.5 to 6.5, we can assume that the pK_a for sulphuric acid is much lower then 5.5. In fact, of course, sulphuric acid is a strong acid, so the concept of pK_a is meaningless as the acid is essentially 100% dissociated.

If all of your answers were correct, return to the main text to continue with Part 3. If not, you may like to revise Sections 3.i and 3.ii.

Self-assessment exercise 3.2

1. Why is the PO_2 in the alveoli lower than the atmospheric value?
   Because the atmospheric air drawn into the respiratory tract becomes mixed with, and in effect diluted by, residual gas which was not fully expired.
2. Why does venous blood have a lower pH than arterial blood?
   Because the PCO_2 is higher (by a little less than 10%) than that in the arterial supply and pH is inversely proportional to PCO_2.
3. The ODC shows that the binding between oxygen and the Hb is allosteric. Carbon monoxide is an irreversible competitive inhibitor of oxygen binding. Sketch a diagram to show the effect CO would have on the oxygen dissociation curve.

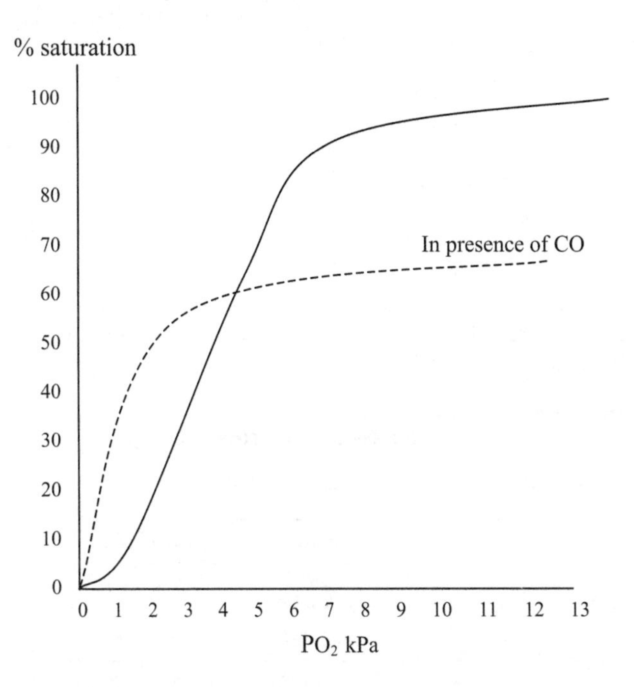

Carbon monoxide binds much more strongly to Hb than does oxygen. In addition, the cooperative allosteric nature of the binding between oxygen and Hb is lost. The net effect is that Hb carries less oxygen and so the tissues become hypoxic. The actual magnitude of the effect depends upon the 'amount' of CO present.

4. What physiological advantage can athletes gain by training at high altitude for a period of time prior to a major sporting event?

   When exposed to lower than normal oxygen tension for a period of time, the body responds by (i) increasing the number of red blood cells released into the circulation (a process stimulated by a peptide called HIF-1), and (ii) the production of 2,3 BPG rises, so the oxygen dissociation curve moves to the right allowing more O_2 release at any given PO_2. In short, the athlete acquires a more efficient oxygen delivery system, so maintaining optimum aerobic metabolism even when PO_2 is low, as will occur during a sporting activity. These adaptive mechanisms are a critical part of the process of acclimatisation to high altitudes.

5. Calculate the ratio of NH_4^+ to NH_3 at pH 6.85 (ICF pH) given that the pK_a for proton dissociation is 9.2.

   $9.2 - 6.85 = 2.35$
   antilog $2.35 = 223.9$.
   ICF pH is lower than the pK_a so most of the compound will be protonated, i.e. NH_4^+

   So the ratio is 224:1 in favour of NH_4^+. This is equivalent to stating that only ~0.45% of the total occurs as NH_3.

These questions cover the material presented in Sections 3.iv to 3.viii. Re-read these sections if you need to clarify any of the information. Or, continue with Section 3.x.

Self-assessment exercise 3.3

1. Case study 1: study the data given below:

	On admission	Reference
pH	7.32	7.35–7.45
PCO_2	8.0 kPa	4.8–5.8 kPa
HCO_3^-	30 mmol/L	23–28 mmol/L

These results are consistent with:
(select ONE answer from those given)
(a) simple respiratory acidaemia
(b) simple respiratory alkalaemia
(c) simple non-respiratory acidaemia
(d) simple non-respiratory alkalaemia
The answer is option (a) simple respiratory acidosis.

The pH is low (acidosis)
PCO_2 is high (respiratory)
$[HCO_3^-]$ is normal (the primary condition is not yet compensated)

2. Case study 2: study the data given below:

	On admission	Reference
pH	7.48	7.35–7.45
PCO_2	3.5 kPa	4.8–5.8 kPa
HCO_3^-	19 mmol/L	23–28 mmol/L

These results are consistent with:
(select ONE answer)
(a) respiratory acidaemia
(b) respiratory alkalaemia
(c) non-respiratory acidaemia
(d) non-respiratory alkalaemia
The correct option is (d) non-respiratory alkalosis.

High pH = alkalaemia
Low PCO_2 indicates hyperventilation
Normal [bicarb] (indicates that this is uncompensated).

3. Predict the type of mixed acid-base disturbance that might occur in someone who has suffered a cardiac arrest.

Cardiac arrest causes a mixed disturbance because the poor/lack of blood perfusion to the tissue will cause cellular hypoxia, and so anaerobic respiration and the production of lactate and other 'metabolic acids' in excess (= metabolic acidosis); the anion gap would increase as bicarbonate concentration falls in relation to the accumulation of metabolic acid, whereas the SID would be low for the same reason. Also we might expect that the removal of CO_2 will be impaired, so there will be a respiratory acidosis.

4. In order to compensate for a respiratory acidaemia, the body would:
   (select ONE answer from those given below)
   (a) Hyperventilate, to expel CO_2
   (b) Hypoventilate to retain CO_2
   (c) Retain more HCO_3^- in the kidney
   (d) Excrete more HCO_3^- via the kidneys
   The correct option is (c), try to retain more bicarbonate in the kidney.
5. Using the Henderson-Hasselbalch equation, confirm validity of the data and then identify the type of acid-base disorder in each case (i) to (vi).

(i) $pH = 7.59$
$[HCO_3^-] = 25\ mmol/L$
$PCO_2 = 3.6\ kPa$

Answer: $pH = 6.1 + \log\{25 \div (3.6 \times 0.225)\} = 7.589$
Data are consistent and likely to be valid.
Respiratory alkalosis (hyperventilation).

(ii) $pH = 7.14$
$[HCO_3^-] = 12\ mmol/L$
$PCO_2 = 4.8\ kPa$

Answer : $pH = 6.1 + \log\{12 \div (4.8 \times 0.225)\} = 7.145$
Data are consistent and likely to be valid.
Metabolic acidosis with some evidence of compensation beginning to operate, i.e. the PCO_2 is at the lower limit of normal.

(iii) $pH = 7.39$
$[HCO_3^-] = 35\ mmol/L$
$PCO_2 = 5.0\ kPa$

Answer : $pH = 6.1 + \log\{35 \div (5.0 \times 0.225)\} = 7.592$
These data are inconsistent and the analysis is invalid.

(iv) $pH = 7.42$
$[HCO_3^-] = 27\ mmol/L$
$PCO_2 = 5.8\ kPa$

Answer : $pH = 6.1 + \log \{27 \div (5.8 \times 0.225)\} = 7.416$
Data are consistent and likely to be valid.
Normal data.

(v)
$$pH = 7.57$$
$$[HCO_3^-] = 14 \text{ mmol/L}$$
$$PCO_2 = 2.1 \text{ kPa}$$

Answer : $pH = 6.1 + \log \{14 \div (2.1 \times 0.225)\} = 7.572$
Data are consistent and likely to be valid.
Respiratory alkalosis, but the HCO_3^- is low and compensation has started, so this must be a chronic condition because the kidneys take 2–3 days to respond in their handling of bicarbonate ion.

(vi)
$$pH = 7.12$$
$$[HCO_3^-] = 9 \text{ mmol/L}$$
$$PCO_2 = 3.8 \text{ kPa}$$

Answer : $pH = 6.1 + \log \{9 \div (3.8 \times 0.225)\} = 7.122$
Data are consistent and likely to be valid.
Metabolic acidosis with some degree of respiratory compensation as shown by the reduced PCO_2.

6. Which of the suggested causes is the most likely to be responsible for the data shown?

Possible causes: diabetic coma
acute renal failure
chronic stimulation of the respiratory centre
barbiturate overdose
chronic respiratory failure

$$pH = 7.20$$
$$[HCO_3^-] = 19 \text{ mmol/L}$$
$$PCO_2 = 8.5 \text{ kPa}$$

This is clearly an acidaemia. The data are inconsistent with any compensatory mechanism having begun to operate. Both HCO_3^- and PCO_2 are abnormal, suggesting either (a) a primary mixed disturbance, or (b) an acute respiratory condition with no renal compensation. Typically, diabetic coma and acute renal failure are associated with metabolic acidaemia, so these can be discounted as options. Chronic stimulation of the respiratory centre would lead to an alkalaemia and assuming normal renal function, chronic respiratory failure would result in increased bicarbonate (compensation). Barbiturate depresses the respiratory centre and so slows carbon dioxide expiration. The consequent rise in $[H_2CO_3]$ causes the low bicarbonate and low pH.

7. The 'ammonium chloride loading test' may be used to assess a subject's ability to acidify her/his urine. What is the physiological basis of this test?

   In an aqueous environment, NH_4Cl would dissociate into $NH_3 + H^+ + Cl^-$ and an acidosis would result. The traditional approach to interpretation based on the Henderson-Hasselbalch equation sees this as an intake of protons which need to be excreted via the kidney, whereas the Stewart model explains this situation as a change in SID due to the increase in $[Cl^-]$, and it is primarily the chloride ions rather than the protons that are excreted.

Were your answers correct? If not, work through the formulae in Sections 3.x to 3.xv again.

Self-assessment exercise 3.4

1. What is the physiological reasoning for the use of saline infusion to treat loss of gastric juice?

   Gastric juice contains HCl, and so vomiting will cause a loss of gastric acid, hence an alkalosis. Loss of ECF (reduced volume) will activate the aldosterone response, which if left uncorrected will exacerbate the alkalosis. Stewart's approach would be to see vomiting as a disproportionate loss of chloride (from the HCl) compared with the loss of sodium causing an increase in the SID and hence an alkalosis, so replacement of Cl is required rather than simply volume expansion with a colloid.

2. What effect would the infusion of a litre of 'normal' physiological saline have on the SID? (*HINT: physiological saline contains only NaCl at a concentration of approximately 150 mmol/L.*)

   Infusion of small volumes of saline will have little effect on SID in the long term. However, if several litres of saline are infused, the relative amount of chloride added is greater than the added sodium because plasma Na is higher than plasma Cl (140 mmol/L and 100 mmol/L respectively). The disproportionate addition of

Cl by the infusion of 150 mmol/L NaCl will reduce the SID and so too the ECF pH.

3. How would the Stewart model explain the acid-base imbalance in pyloric stenosis?

   Stewart's approach would be to see this as a disproportionate loss of chloride compared with the loss of sodium causing an increase in the SID, and hence an alkalosis.

Appendix

Appendix I: Glossary and abbreviations

Term	Definition
Acid	According to the Lowry-Bronsted theory, an acid is a compound that dissociates (ionises) in water to liberate protons (hydrogen ions; H^+). Strong acids dissociate completely whereas weak acids dissociate only very slightly (see Weak electrolytes below).
Acidaemia	An abnormal physiological state that results in the pH of arterial blood being less than 7.36. This indicates the presence of an acidosis.
Acidosis	An underlying pathological condition that is likely to cause a fall in arterial blood pH to less than 7.36; it is not necessary for the blood pH actually to be below the reference limit.
Alkalaemia	An abnormal physiological state that results in the pH of arterial blood being greater than 7.44. This in effect usually indicates the presence of an alkalosis.
Alkalosis	An underlying pathological condition that is likely to cause a rise in arterial blood pH to greater than 7.44; it is not necessary for the blood pH actually to be above the reference limit.
Anion	A chemical species that carries a negative charge.
Anion gap (AG)	The numerical difference between the concentrations of measured anions and measured cations in plasma: $$AG = (Na + K) - (HCO_3 + Cl).$$ Physiologically, the AG does not exist as body fluids exhibit electrical neutrality, but the AG is often useful to identify the presence in plasma of 'abnormal' anions in certain cases of acidosis. The corrected anion gap includes an allowance for changes in albumin concentration.

(*Continued*)

Essential Fluid, Electrolyte and pH Homeostasis, First Edition. Gillian Cockerill and Stephen Reed.

Term	Definition
Base	According to the Lowry-Bronsted theory, a base is a compound that accepts protons (H^+), thus reducing the hydrogen ion concentration. Strong bases accept protons strongly; weak bases accept protons weakly.
Base excess or base deficit	Base excess is an estimate of the magnitude of any metabolic derangement of acid-base status independent of any respiratory component. A base excess indicates an alkalosis and a base deficit (or negative base excess) indicates an acidosis. The value expresses the amount of acid or alkali required to return blood to pH 7.40 at a normal PCO_2 and normal temperature.
Buffer	A solution that resists a dramatic change in pH. Physiologically important buffer solutions are composed of a weak acid and a base (salt) of that same weak acid.
Buffer base	A collective term that attempts to quantify the total concentration of non-bicarbonate buffers in blood.
Cation	A chemical species that carries a positive charge.
ECF	Extracellular fluid.
Electrolyte	A chemical species with a net overall charge. See Anions and Cations.
Exudate	A fluid that seeps out of the bloodstream usually at the site of inflammation. Examples include ascitic fluid and plural fluid. See Transudate.
Homeostasis	A physiological state of chemical balance (or 'normality'). Homeostatic control or mechanisms bring about or maintain that state of balance.
ICF	Intracellular fluid.
Osmolality	A function of the total solute concentration of a fluid, osmolality is an estimate of tonicity, the true physiological 'force' that tends to control fluid movement between compartments. Osmolality may be easily measured in plasma or urine in the clinical laboratory. Units of measurement are mmol/kg (or mOsm/kg). See also Osmolarity and Tonicity.
Osmolarity	This is a surrogate measure of osmolality. Osmolarity is a calculated parameter based on the concentration of the principal osmotically active species (osmolytes) present in the body fluid. The unit of measurement is mmol/L. The true physiological osmolality is usually about 10–15 units higher than the calculated osmolarity.
PCO_2	Partial pressure of carbon dioxide; an estimate of CO_2 gas 'concentration' or 'amount' in blood.
PO_2	Partial pressure of oxygen; an estimate of O_2 gas 'concentration' or 'amount' in blood.

Term	Definition
Standard bicarbonate	This is a measure of the 'metabolic component' of an acid base disturbance. The value is the theoretical value of arterial plasma bicarbonate assuming the PCO_2, temperature and oxygen saturation to be normal.
Strong electrolyte e.g. a strong acid	One which is fully dissociated: $$BZ \longrightarrow B^- + Z^+$$ for a strong acid; $$BH \longrightarrow B^- + H^+$$ Unlike the situation with weak electrolytes and weak acids, there is no equilibrium.
Strong ion difference (SID)	The SID is a parameter derived by Peter Stewart in his so-called 'quantitative' approach to understanding acid-base disturbances. Numerically, SID is a measure of unmeasured strong anions such as lactate. Compare SID with anion gap.
Tonicity	The effective osmotic potential of a solution. Where two solutions are separated by a selectively permeable membrane, the net flow of liquid from one fluid compartment to the other is determined by the relative concentrations of *non-diffusible* chemical species present in the solutions. Relative tonicity of body fluids is the 'driving factor' affecting fluid movement between interstitial fluid and intracellular fluid. Compare tonicity with Osmolality and Osmolarity. Note that urea is a diffusible solute (except in the renal tubule) so it has little or no impact on osmotic pressure across cell membranes, but urea is a contributor to measured osmolality.
Transudate	A fluid which passes by a process of ultrafiltration from the blood stream into a tissue space. Examples include CSF and saliva.
Weak electrolyte; e.g. a weak acid	One which is not fully dissociated. A chemical equilibrium is established between the ionised and un-dissociated forms with the relative concentrations favouring the un-dissociated form: $$AX \rightleftharpoons A^- + X^+$$ for a weak acid; $$AH \rightleftharpoons A^- + H^+$$ The position of equilibrium lies heavily to the left and so the concentration of free protons (H^+) is very small.

Appendix II: Reference ranges

Reference ranges for commonly measured analytes are to be used as a guide for interpretation only, as actual values quoted may vary between laboratories.

Analyte	Range
Plasma	
Anion gap	11–16 mmol/L or 15–20 mmol/L (depending on formula used)
Bicarbonate	23–28 mmol/L
Calcium (total)	2.25–2.55 mmol/L
Chloride	97–107 mmol/L
Glucose (fasting)	3.5–5.5 mmol/L
Magnesium	0.8–1.1 mmol/L
Osmolality	285–295 mmol/kg
Potassium	3.5–4.5 mmol/L
Sodium	135–145 mmol/L
Urea	3.0–7.0 mmol/L
Whole arterial blood	
pH	7.35–7.45
PCO_2	4.8–5.8 kPa
PO_2	12–14 kPa
Base excess	−2 to +2 mmol/L
Standard bicarbonate	22–26 mmol/L
Urine output per 24 hours	
Sodium	100–200 mmol
Potassium	50–100 mmol
Calcium	4.5–6 mmol
Creatinine	8–12 mmol
Volume	Variable according to fluid intake as liquids

Index

f signifies a Figure *t* signifies a Table

Essential Fluid, Electrolyte and pH Homeostasis, First Edition. Gillian Cockerill and Stephen Reed.